T0293501

Current Trends in Earth Observation: Technology and Innovation

Current Trends in Earth Observation: Technology and Innovation

Editor: Kallie Morgan

www.callistoreference.com

Callisto Reference,
118-35 Queens Blvd., Suite 400,
Forest Hills, NY 11375, USA

Visit us on the World Wide Web at:
www.callistoreference.com

© Callisto Reference, 2023

ISBN: 978-1-64116-783-3 (Hardback)

Cataloging-in-publication Data

Current trends in earth observation : technology and
innovation / edited by Kallie Morgan.
    p. cm.
Includes bibliographical references and index.
ISBN 978-1-64116-783-3
1. Earth sciences--Remote sensing. 2. Earth (Planet)--Observations.
3. Earth sciences. I. Morgan, Kallie.
QE33.2.R4 C87 2023
550.28--dc23

Table of Contents

Preface

I am honored to present to you this unique book which encompasses the most up-to-date data in the field. I was extremely pleased to get this opportunity of editing the work of experts from across the globe. I have also written papers in this field and researched the various aspects revolving around the progress of the discipline. I have tried to unify my knowledge along with that of stalwarts from every corner of the world, to produce a text which not only benefits the readers but also facilitates the growth of the field.

Earth observation refers to the gathering of information about the physical, chemical and biological systems of the planet Earth. It involves monitoring and assessing the changes in the natural and manmade environments. Earth observation is performed via remote sensing satellites and various in-situ instruments. There are various types of Earth observation instruments such as radars and sonars, thermometers, wind gauges, ocean buoys, and seismometers. Different types of Earth observations include photographs, radar and sonar images, analyses of water and soil samples, and processed information such as forecasts, maps and models. The major applications of Earth observations are forecasting weather; predicting, adapting to and mitigating climate change; and addressing emerging diseases and other health risks. Apart from these, tracking biodiversity and wildlife trends, and measuring land-use change are also important applications of Earth observations. This book includes some of the vital pieces of work being conducted across the world, on the technology and innovation related to Earth observation. It presents researches and studies performed by experts across the globe. Through this book, an attempt has been made to further enlighten the readers about the new concepts in this area of study.

Finally, I would like to thank all the contributing authors for their valuable time and contributions. This book would not have been possible without their efforts. I would also like to thank my friends and family for their constant support.

Editor

Artificial Intelligence and Earth Observation to Explore Water Quality in the Wadden Sea

Luigi Ceccaroni, Filip Velickovski, Meinte Blaas, Marcel R. Wernand, Anouk Blauw, and Laia Subirats

Abstract Earth-observation systems (satellites and in situ monitoring) are routinely used to collect information about water quality. Recently, smartphone-based tools and other citizen-science sensors have enabled citizens to also contribute to the collection of scientifically relevant data. This chapter describes a decision support system used to predict optical water-quality indicators in the Wadden Sea, which is an intertidal marine system, where natural processes related to sediment transport and primary production define the basis of its ecological values. As information sources, the system uses satellite data, data collected with a mobile app and physical data for the period 2003–2015. An artificial-intelligence technique, inductive learning, is used to analyze the data and provide predictions in terms of water colour represented via the Forel-Ule scale (a comparative scale for colour).

Introduction

The Wadden Sea is a large-scale, intertidal marine system, where natural processes related to sediment transport and primary production define the basis of its internationally recognized ecological values. Human pressures on the system abound: aquaculture, fishing, tourism and recreation, mining activities, agriculture and industry in the surrounding region. Nutrient inputs from rivers, growing tourism and large-scale fishery activities may create conditions that negatively affect human

L. Ceccaroni (✉)
1000001 Labs, Barcelona, Spain
e-mail: luigi@1000001labs.org

F. Velickovski • L. Subirats
Eurecat, Barcelona, Spain
e-mail: filip.velickovski@eurecat.org; laia.subirats@eurecat.org

M. Blaas • A. Blauw
Deltares, Delft, The Netherlands
e-mail: Meinte.Blaas@deltares.nl; Anouk.Blauw@deltares.nl

M.R. Wernand
Royal Netherlands Institute for Sea Research, Texel, The Netherlands
e-mail: Marcel.Wernand@nioz.nl

society, water quality and ecological systems. Satellites and in situ monitoring are routinely used to collect information about water quality (see Fig. 1). Recently, smartphone-based tools and other citizen-science sensors have entered the arena to enable citizens to collect scientifically relevant data (Graham et al. 2011; Ceccaroni and Piera 2017), which can be used by decision support systems. This chapter describes the Citclops Data Explorer: a knowledge-based system designed to predict optical water-quality indicators in the Wadden Sea that may be used for aquaculture, tourism, recreational diving, and water management. As *information sources*, the system uses MERIS satellite data, data collected with the *Citclops—Citizen water monitoring* app (Wernand et al. 2012) and physical data: waves, currents, river inputs and weather data for the period 2003–2015 (MERIS data available up to 2011; app data available in 2014–2015).

Inductive Learning

The inputs to the system were a vector of numerical attribute values, and the target value was a discrete integer. Inductive learning was used to analyse the input data and provide predictions in terms of *water colour,* using the *Forel-Ule* (FU) scale, a comparative scale developed in the nineteenth century. The FU scale has an implicit relation to other water-quality properties such as turbidity, transparency, suspended particulate matter and chlorophyll (Wernand 2011). In this way, it is possible to learn a general function or rule from a specific set of input-target value pairs. The system used an artificial-intelligence technique, semi-supervised learning, for

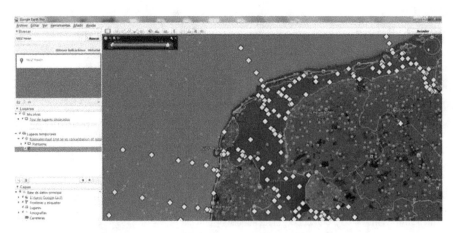

Fig. 1 In situ monitoring platform: *yellow markers* indicate the Dutch national water quality monitoring network (Rijkswaterstaat); the *red pin* (NIOZ jetty) indicates the location of the observation platform of the Royal Netherlands Institute of Sea Research (NIOZ). Source: http://kml.deltares.nl/kml/rijkswaterstaat/waterbase/concentration_of_suspended_matter_in_water.kml and Google Earth

capturing the model that establishes the relationship between input and target value pairs. Part of the water colour data was collected by ordinary citizens snapping pictures and being asked to select the most appropriate colour via the *Citclops—Citizen water monitoring* app. As this is a citizen-science setting, a degree of noise and inaccuracies are expected and dealt with via quality control techniques that involved: the automatic analysis of the photo image, comparison against known satellite measured colour, and flagging of inappropriate measurements via citizen peers.

A system with *the ability to predict water quality* can be useful in several applications. Apart from direct uses in recreation apps by citizens, it can assist water managers in long-term monitoring, system analysis and decision making on water use. It can provide information to assess the constraints and opportunities for sustainable use of the sea and coast, and also guide risk analysis and response to early warnings. With the information sources mentioned above, an inductive *machine learning technique* (decision trees) was employed to predict water colour 1 week in the future.

The design of the learning technique took into account three major issues: (1) the output or target value of the model to be learned; (2) the *feedback* available to system; (3) the *representation* of the learned model. The target *value* to be learned was water colour. The *type of feedback* available determined the nature of the learning problem that the system faced: *semi-supervised* learning, which involves learning a *function* from examples of inputs and outputs. The system learned a model represented as a function that maps observations of MERIS satellite data, citizen data and physical data to a discrete output (colour represented as FU). Finally, the *representation of the learned information* was a decision tree determined by the type of learning algorithms being used. The last major factor in the design of the learning system was the *availability of prior knowledge*. The system began with no knowledge at all apart from the examples in the data series.

In this study, a machine learning framework is described that uses semi-supervised learning to generate a predictive model that maps marine data coming from heterogeneous sources to a water quality indicator: colour represented by FU. Decision tree induction was used, being one of the most successful forms of learning algorithms, and the model generated is explicit and natural for human data-interpretation. *Decision trees* take as input a situation described by a set of attributes (from remote sensing, citizens and in situ instruments) and return a *decision*: the predicted output value for the input, i.e. the prediction of the evolution of FU colour 1 week ahead in time. The input attributes are continuous. The target value is a fixed set of values; therefore the problem can be constructed as a *classification* learning problem.

Decision trees classify the input vector by performing a sequence of tests. Each internal node in a tree corresponds to a test of the value of one of the attributes in the vector, and the branches from the node are labelled with the possible values of the test. Each leaf node in a tree specifies the value to be returned.

The aim here is to learn a model for the target label *FU-Colour*.

Data Description

The following list describes the data available to the system:

1. MERIS satellite data: FU, chlorophyll-a (2002–2011)—time resolution: one data point per day (missing data on cloudy days)
2. FU data collected with the *Citclops—Citizen water monitoring* app (2013–2015)
3. Water quality data: *total suspended matter* (TSM), FU, chlorophyll-a collected by means of a spectroradiometer mounted on the NIOZ observation platform (2013–2015) (see Figs. 2 and 3)—time resolution: every 2 min during daylight (some missing periods)

Fig. 2 The study area for the 2013–2015 study period is the Wadden Sea, located within the rectangle at the center of the figure. Source: Google Earth

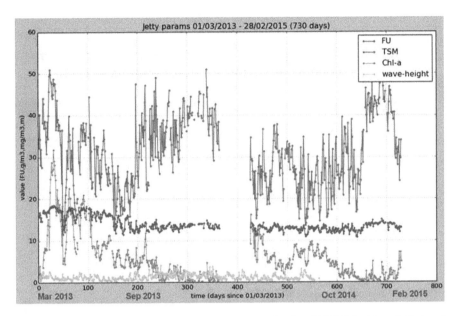

Fig. 3 NIOZ jetty Wadden Sea: RS calculated water quality data (TSM, FU, chlorophyll-a) and wave height (daily means)

4. Wave data (2003–2013) (wave height and period)—average time resolution: one data point per hour
5. Water-current data (2003–2013)
6. River inputs, salinity, water temperature (2003–2013)—average time resolution: one data point per day
7. Weather data: insolation, as an extra driver for algal growth; wind speed magnitude, which correlates strongly with wave height; wind direction; and air temperatures (2003–2013)
8. SPM, chlorophyll-a, DOC, K_d collected in situ (2003–2013)—average time resolution: two data points per month

Methods

The following Earth observations have been finally used as part of the input: FU colour index, TSM, chlorophyll-a and wave height. A model of the target variable "FU colour" at future points (2 days, 4 days, 7 days) has been learned (see Fig. 4). To do this, the initial problem was converted to a three-class classification problem (see Fig. 5): FU colour *decreases*, is *stable* or *increases*.

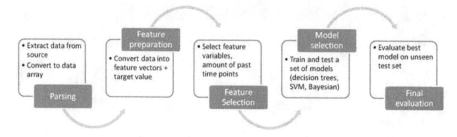

Fig. 4 Machine-learning pipeline

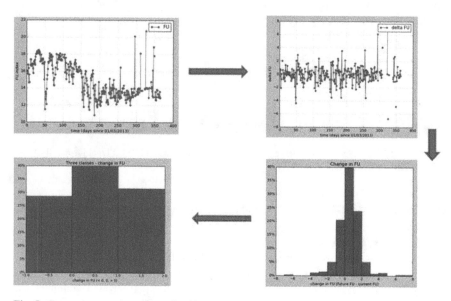

Fig. 5 Conversion to a three-class classification problem

The model's prediction of FU has been evaluated using tenfold cross-validation. It has then been integrated into the Citclops Data Explorer—Marine Data Analyser (http://citclops-data-explorer.herokuapp.com/marine-data-analyser).

Results and Discussion

Note that every variable used has a small set of possible values or is continuous; the value of *FU colour index*, for example, is not an integer, rather it is one of the 21 discrete values from 1 to 21. The task of finding a tree that is consistent with the input examples and is as small as possible, no matter how size is measured, is an intractable problem: time grows exponentially with the amount of data and there is no way to efficiently search through the possible trees. With some simple

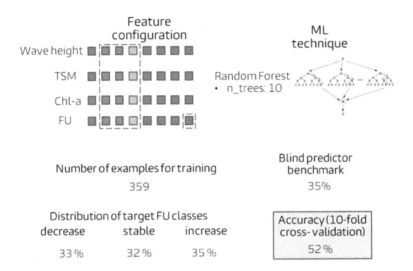

Fig. 6 Example of results using a specific feature-configuration and a support vector machine algorithm

heuristics, however, the authors found a good approximate solution: a small (but not the smallest) consistent tree, defining the sequence of tests and the specification of each test in an acceptable time.

The forecasting system is composed of different decision trees (implemented in Python), which predict if the FU colour *decreases*, is *stable* or *increases* over a week (2 days, 4 days and 7 days in advance). The performance of these decision trees is compared to the one of a support vector machine algorithm and of blind predictors. Samples of the results of the algorithms used are presented in Figs. 6, 7 and 8.

Each figure represents the learning protocol and experiment that was performed. The rows of the grid on the top left of the figure are the type of attributes (wave height, TSM, Chl-a, FU), and the columns are consecutive individual days on which the attributes have been measured. The coloured (non-blue) squares mark the feature configuration of the training examples. The column with the squares coloured in orange represent the reference time point ($t = 0$/present time point) and are part of the input vector. The red squares are attributes that are also included in the input vector but from days before the reference time point. The green square is the attribute that the model will learn to predict which will always be at a future time point in relation with the orange column. In the top right is the learning technique and some key configuration parameters. As an example, in Fig. 6, the target-value attribute is FU at 2 days into the future, and the input vector includes the following attributes: wave height, TSM, Chl-a, FU at the current time point, and wave height at 1 day in the past.

The algorithm used adopts a greedy divide-and-conquer strategy: always test the most important attribute first. This test divides the problem up into smaller sub-problems that can then be solved recursively. By "most important attribute", the

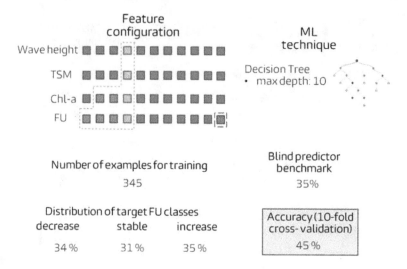

Fig. 7 Example of results using a specific feature-configuration and random-forest decision trees

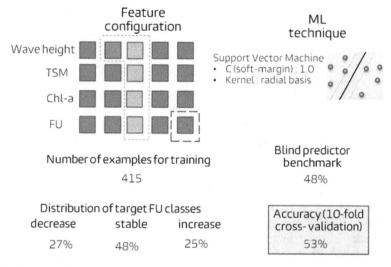

Fig. 8 Example of results using a specific feature-configuration and decision trees with a maximum depth of 10

authors mean the one that makes the most difference to the classification of an example. That way, the authors hope to get to the correct classification with a small number of tests, meaning that all paths in the tree will be short and the tree as a whole will be shallow.

In general, after the first attribute-test splits up the examples, each outcome is a new decision-tree learning problem in itself, with fewer examples and one less attribute. There are four cases to consider for this recursive problem:

1. If the remaining examples are all *decrease* (or *stable* or *increase*), then the algorithm provides an answer.
2. If there are some mixed decrease, stable or increase examples, then choose the best attribute to split them.
3. If there are no examples left, it means that no example has been observed for this combination of attribute values, and the algorithm returns a default value calculated from the plurality classification of all the examples that were used in constructing the node's parent.
4. If there are no attributes left, but both positive and negative examples, it means that these examples have exactly the same description, but different classifications. This can happen because there is an error or *noise* in the data; because the domain is nondeterministic; or because an attribute that would distinguish the examples has not been observed or taken into account. The algorithm returns in this case the plurality classification of the remaining examples.

The accuracy of the learning protocol is compared in each case to a blind predictor as a benchmark test. The blind predictor always classifies to the most common class in the examples of the training set. In the case of Fig. 8, the most common class is an increase in FU, that occurs 35% of the time. Thus a classifier predicting always an increase would be 35% of time accurate. The accuracy by the decision-tree algorithm is 45% thus suggesting that indeed the model has utilised patterns in the current attributes and past attributes to predict the future value (7 days ahead in time).

Conclusions

In this study, an artificial-intelligence technique, inductive learning, has been used to analyze data from Earth-observation systems, citizens, marine scientists and coastal planners and to provide predictions in terms of water colour, using the Forel-Ule scale, a comparative scale for colour. Specifically, decision trees have been used for learning. Note that the set of data examples is crucial for *constructing* the trees, therefore the quality of the trees as a classification tool depends on the quality of the original data. Each tree consists of just tests on attributes in the interior nodes, values of attributes on the branches, and output values on the leaf nodes.

These trees are also bound to make some mistakes for cases where they have seen no examples. For example, they have never seen cases of extreme FU values. In future work, with more training examples, the learning program could correct these mistakes.

The authors could identify the following potential applications for these trees:

- to provide sea farmers with bulletins about algal blooms, which change the water colour;
- to maximize citizens' experience in activities in which water quality has a role; and

- to provide citizens with powerful, user-friendly tools for environmental-data interpretation.

References

Ceccaroni L, Piera J (2017) Analyzing the role of citizen science in modern research. IGI Global, Hershey, PA. https://doi.org/10.4018/978-1-5225-0962-2

Graham EA, Henderson S, Schloss A (2011) Using mobile phones to engage citizen scientists in research. EOS Trans Am Geophys Union 92(38):313–315

Wernand MR (2011) Poseidon's paintbox: historical archives of ocean colour in global-change perspective. Ph.D. thesis, Utrecht University, p 240. ISSN 978-90-6464-509-9

Wernand MR, Ceccaroni L, Piera J, Zielinski O (2012) Crowdsourcing technologies for the monitoring of the colour, transparency and fluorescence of the sea. In: Proceedings of ocean optics XXI, Glasgow, Scotland, pp 8–12

2

Sustainable Agriculture and Smart Farming

Heike Bach and Wolfram Mauser

Tomorrow's challenges of doubling food supply put sustainability of agriculture at one level with ensuring food security. The global food system needs to be resource efficient and at the same time sustainable. Efficient use of water, reduction of soil erosion and degradation to the minimum, minimization of energy input and maximization of yields under uncertain natural conditions are the goal. They pose highest requirements on the underlying information and knowledge infrastructure and make future farming a knowledge business and a very sophisticated management task.

Studies on the global and regional potentials of Earth Observation (EO) in agriculture show that EO can be pivotal. Due to its global capacity to determine information relevant for farming EO can become the global source of future, information driven global agriculture (Mauser et al. 2012). Assimilated into sophisticated environmental management models this EO-derived information will allow to support the whole economic and societal value chain from farmers through food industry to insurance and financial industry in producing food. At the same time, it allows to support society in governing sustainable agriculture through verifiable rules and regulations.

Information driven smart farming is a general trend to be observed in agriculture. Ecologically and economically meaningful measures to improve productivity are applied in smart farming. The technique is based on the principles of Precision Farming, i.e. on the use of GPS-guidance to apply site-specific agricultural measures. But while the focus of Precision Farming was mainly on farming technology to for example allow for auto-steering of tractors and harvesters, the focus of smart

H. Bach (✉)
Remote Sensing in Geosciences, VISTA GmbH, Munich, Germany
e-mail: bach@vista-geo.de

W. Mauser
Department of Geography, University of Munich LMU, Munich, Germany
e-mail: w.mauser@lmu.de

farming shifts towards a more rounded, holistic approach—going from "highest spatial precision" to "smartest treatment". Thus, typical issues of smart farming are e.g. how much fertilizer is best applied when and where in the field or which plant protection resources are optimal for crop development at each location in the field.

The information challenge agriculture is facing is manifold. High spatial and temporal requirements are posed on a monitoring system since the plots where food is produced are in general quite small. Their size varies largely depending on economic and cultural conditions, but 10–20 m can be deemed the most suitable spatial observational requirement for an agricultural information system, which satisfies foreseeable future needs. This required spatial resolution also fits with the capabilities and spatial accuracies of site-specific farming, which is determined by the working width of the agricultural machinery: seeders (5–10 m), spreaders and combine harvesters (20–40 m). The dynamic growth of agricultural crops, and man-made changes within a few days for example through harvests further make it necessary to update the information flow every few days to one week. However maybe most challenging are the complex information requirements since complex information layers like yield or nitrogen uptake are needed. They are in general no direct EO observable.

Sustainable agriculture and smart farming need data driven information services. These support sustainable and cost-effective agriculture by combining Earth Observation and navigation satellites' input with information from ground sensors to help farmers decide how, when and where to allocate resources for the best economic and ecological results (Fig. 1). As use case showing how this is presently applied in farming practise, the TalkingFields smart farming services will be presented (Bach et al. 2010).

EO has been used for agriculture since the 80th of the last century, however the availability and quality of the satellite data has increased drastically since 2015 with the Copernicus program and its operational feet of satellites. Especially Sentinel-2 and -1 are providing excellent time series of data from which e.g. crop types, biomass development, calamities and farming practices (ploughing, seeding, etc.) can be derived with high accuracy.

The free availability of Landsat's complete data archive that is now supported with Sentinel-2 allows for new data analyses techniques. They make use of very data intensive exploration and data fusion techniques. It is now possible to analyse dozens of satellite images of one farm in order to understand the site-characteristics of each field of the farm and even within the fields. The TalkingFields (TF) Base Map illustrated in Fig. 2 is a good example for this procedure.

Multi-Year Site-Characterization of Fields Using Advanced Data Harvesting Techniques

The TF Base Map is based on a geo-statistical analysis of multi-year optical data to map the spatial heterogeneity of the growing conditions within the field. It uses all available satellite images of the last 5–10 years and evaluates specific multiannual features that can be expressed as relative fertility. Often more than 100 scenes are processed in order to get the best representation of site heterogeneity. Pattern recognition techniques also allow for improved segmentation of the field. The TF Base Map can then be used for improved sampling of soil properties or delineating of management zones. The TF Zone Map optimizes the segmentation for soil sampling tasks. Based on the whole variety of in-field growth conditions, it delineates homogeneous regions and thereby ensures that soil samples are representative, as illustrated in the centre field in Fig. 2.

The TF Base Map opens to the farmer a new view on his fields by displaying the in-field differences much in the same way that crops experience when they turn his inputs into yield. The patterns observed in the Base Map are usually bound to site characteristics, e.g. the water holding capacity of the soil based on differences in soil properties or relief related water status. Even within one single field the small-scale changes of yield formation can easily vary by $\pm 25\%$. This is often not known to the farmer since conventionally yield data are obtained field wise. Thus these EO based maps are a prerequisite for smart farming in the sense that they inform the farmer of the heterogeneity within his fields.

Besides the trend of moving from analysis of single satellite images to multi-temporal stacks of images, also data assimilation has become more prominent. Data assimilation of EO data into process models is already a common practise for weather forecast. METEOSAT data are successfully used in the Numerical Weather Prediction Models to make the forecasts more reliable. This concept can now be

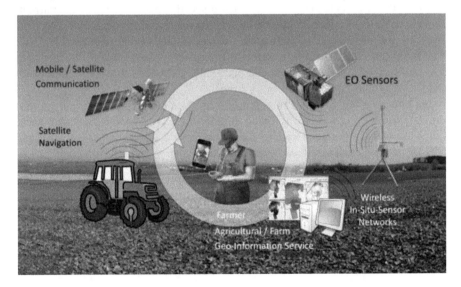

Fig. 1 Information flow in TalkingFields. Space-based components play an essential role in smart farming information services for farmers

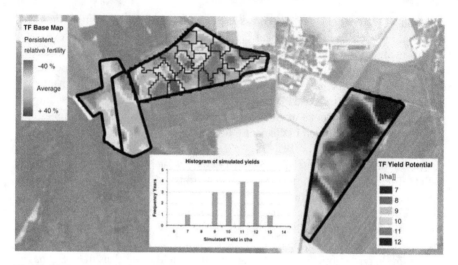

Fig. 2 EO based map products used in smart farming; the TF Base Map offers the persistent relative fertility of a field by analysing satellite images of the last 5–10 years; these TF Base Maps can be segmented in zones for improved (more representative) soil probing; after combining the TF Basemap with crop growth modelling even maps of yield potential can be derived

transferred to high resolution and the land surface generating new products. This is for example the case when extending the information product TF Base Map to a Yield Potential Map.

The yield potential is a quantitative term expressed in t/ha and describes how much yield can be obtained for a field and a specific crop type under the current climatic and topographic conditions assuming no nutrient stress nor pests or diseases. The yield potential is thus also describing the optimum attainable yield if all farming decisions are made correctly. In reality this level is hardly reached and large differences exits in the yield gap that can be calculated from the actual yield and the yield potential.

For obtaining the yield potential the TF Base Map is merged with crop growth simulations of the last 16 years. Using the PROMET crop growth model for each simulated year the potential yield is simulated (for details see also (Mauser et al. 2015)) and the annual results are averaged for the map product illustrated in Fig. 2. In order to also learn about the exposure of the field to climate risks, e.g. droughts, also a histogram analysis of all simulated years is included in the analyses. In this example the variance of yield potential is quite large. In dry years, only 7 t/ha wheat can be harvested, whereas even 13 t/ha can be achieved in years with sufficient rainfall. Water is obviously the dominant factor of yield formation in this German region under consideration and accordingly weather risks must be considered rather high.

Up-to-Date Crop Status Using Access to Satellite Data in Near-Real-Time

Another trend in Earth Observation is the faster availability of satellite information products. Not only EO data providers offer their data sets in near-real-time, also agricultural information services are now able to generate up-to-date information on crop status within 24 h after data acquisitions. Information about the current growth status and development of the cultivated crop at each location in the field is deduced from satellite through the derivation of plant parameters such as LAI, biomass and chlorophyll content during the growing season. The up-to-date crop status is used for example for site-specific plant protection and fertilization measures.

TalkingFields also offers this up-to-date information to farmers e.g. with the TF Biomass Map. This map product represents the above ground biomass distribution for a given date. A sequence of Biomass Maps derived from a time series of satellite images documents the temporal and spatial patterns of the biomass development in the observed field. Data assimilation techniques are again needed to enhance the value of satellite data. Accordingly, the TF Yield Maps offer yield information that are determined by combining a plant growth model with time series of multi-sensor EO data. They can be provided already weeks before harvest and validated with at-harvest measurements by a combine.

On the top of Fig. 3 three Biomass Maps can be seen that were assimilated into the agricultural model PROMET to update the daily biomass simulations and nudge the model towards the observed patterns. On the bottom, the resulting yield map can be seen (left side) and compared with the yield map as measured by a GPS-guided combine harvester (right side). It is clearly visible that both yield maps show the same spatial patterns as well as comparable absolute yield values. Since yield is the deciding factor of the farmer's income, a solid validation of the absolute yield values is necessary and was successfully conducted for several years and crop types in Germany (Hank et al. 2015). The yield forecast maps are delivered between 2 and 4 weeks before harvest and give the farmer an overview over the amount of yield expected. If combined with a map of the ripening status, it can also be used to decide where to start with the harvest improving the logistics of harvest machinery.

On-farm research techniques can be used for validating the efficiency and success of precision farming techniques in a realistic scenario (Migdall et al. 2013). On-farm research results of TalkingFields activities for example showed that reduction of nitrogen input can even coincide with an increase in yield. Yield increase of 3–6% could be achieved even on best soils providing a net profit gain to the farmer between 60 and 120 €/ha.

Especially Sentinel-2 strongly contributes to these up-to-date agricultural products and related services. Fast and easy access to the satellite data are offered via dedicated data hubs, like ESA's Sentinel-2 hub, or national platforms, like for example CODE-DE as German service offer. Private companies like Amazon offer not only to download data but also allow further processing directly on the platform. Sponsored by ESA, dedicated Thematic Exploitation Platforms are under

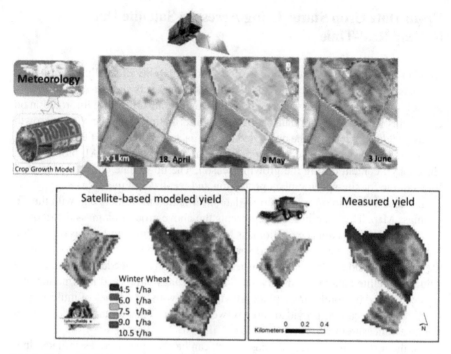

Fig. 3 Data assimilation concept for generating agricultural information services; the Talking-Fields Yield Map obtained using the PROMET crop growth model corresponds very well with measured yields (Hank et al. 2015)

development that not only bring the processing capacities to the data but also provide thematic dedicated analysis tools. For agricultural applications, the Food Security TEP is presently under development to fulfil this goal. The trend for fast data and information access will thus continue.

Considering the required temporal resolution, a weekly sampling interval during the vegetation period can be considered optimum for agricultural applications, the phenological development largely being covered by this frequency. However, depending on the cloud probability of the respective region, this repetition rate is hardly ever available even when combining several sensor systems. Thus, the time gaps between the remotely sensed images are bridged with the crop model simulations which also lower the minimum demand for satellite images per growing season. For yield estimates a set of four images, covering the major stages of development, is considered appropriate (Hank et al. 2013). For site-specific fertilization and harvest logistics weekly updates from satellite are still mandatory. Accordingly, the twin constellation of Sentinel-2 is important to meet the temporal requirements in agriculture.

The vast amount of satellite data to be processed, further make it necessary that crop parameter retrieval methods are automated, although still validated in an unsupervised way, generally applicable and thus geographical transferable. Simple regressions based on spectral indices cannot fulfil this request. Accordingly, in TalkingFields the spectral information used to derive plant physiological parameters is interpreted through inversion of radiative transfer models that are based on the physical modelling of absorptions, transmissions and scattering. Especially leaf area, APAR and chlorophyll content are importance spectrally derivable crop variables, for which an adequate coverage of the visible and near infrared region is vital. Sensitivity studies with the radiative transfer model SLC (Soil-Leaf-Canopy) (Verhoef & Bach 2007) have shown that spectral bands in the short-wave infrared substantially increases the accuracy of leaf area retrieval. Also, for a more detailed analysis, e.g. of the moisture status of the crops using multi-spectral data, a spectral extension to the short wave infrared domain is necessary.

Conclusions and Outlook

EO based services and related products have been proven to be a stable source of information with reliable quality under varying soil and weather conditions. Information products, like the TalkingFields maps, allow the farmer to more accurately react with site-specific farming techniques. More accuracy means lower production costs, as resources such as water, seeds and fertilizer are not wasted. More accurate also means more efficient in the sense of yield per fertilizer or water used. Last but not least, more accurate also means more sustainable because less fertilizer is leached to the ground water. This benefits on the one side farmers with increased economic profits. On the other side, it also serves natural resource providers such as drinking water suppliers or environmental protection agencies, resulting in both successful commercial business and as well as environmental gains. All in all, smart farming supports ecologically and economically sound agricultural management via site-specific applications – an important step towards sustainable agriculture.

For a wide acceptance, it is however needed to offer these services in an easy and integrative way, e.g. within the farmer's Farm Management Information System (i.e. the software used by the farmers to plan and document the activities on their fields). This integration also allows creating application maps for plant protection or fertilization. These application maps can be exported to the farming machinery and be directly used in combination with the GNSS-guided tractor. This integrative approach that covers the whole service chain from the satellites to the tractor terminal will largely promote the acceptance of the services by the farmers and increase their impact on sustainability.

Sentinel-2 already marks a huge step towards better spectral coverage and operational availability of satellite data. The free and open access to Sentinel-2 data and infrastructure platforms like FS-TEP will guarantee that these data can be used by everybody around the Globe, no matter whether they are from an industrialized country or a developing country. This gives equal chances to everybody to be more sustainable in agriculture.

Future sensor development towards hyperspectral coverage will enable to more accurately derive the currently used crop variables and extend the range of parameters available to the crop growth model to further increase its accuracy especially in optimizing crop growth management (Migdall et al. 2012). This will create major improvements in the ability to fine-tune agricultural management towards sustainable yet intensive agricultural production.

Further research and development is still needed in order to expand EO capabilities and services towards comprehensive sustainable farm management, which ensures efficient, water saving irrigation, fertilizer saving fertilization, robust, limited and timely plant protection, stable vegetation cover to minimize erosion and both high quantity and high quality yield. The tightening water-food-energy nexus makes clear that EO-based sustainable farm management is a multi-parameter task. Accordingly addressing future EO mission concepts, science as well as application will need constellations of dedicated sensors, which work in synergy and feed sophisticated land surface process models, which, like in meteorology, deliver products of value for society.

References

Bach H, Migdall S, Mauser W, Angermair W, Sephton AJ, Martin-de-Mercado G (2010) An integrative approach of using satellite-based information for Precision farming: TalkingFields. Proceedings 61st International Astronautical Congress, Prague, CZ

Hank T, Frank T, Bach H, Mauser W (2013) On the effect of multiseasonal Earth Observation availability for the assimilation-supported modelling of winter wheat, Proceedings ESA Living Planet Symposium 2013, Special Publication SP-722, Edinburgh, UK

Hank T, Bach H, Mauser W (2015) Using a remote sensing supported hydro-agroecological model for field-scale simulation of heterogeneous crop growth and yield: application for wheat in Central Europe. Remote Sens 7:3934–3965. https://doi.org/10.3390/rs70403934

Mauser W, Bach H, Hank T, Zabel F, Putzenlechner B (2012) How spectroscopy from space will support world agriculture. IGARSS2012 Munich, IEEE 2012 International Geoscience and Remote Sensing Symposium Proceedings, pp 7321–7324

Mauser W, Klepper G, Zabel F, Delzeit R, Hank T, Calzadilla A (2015) Global biomass production potentials exceed expected future demand without the need for cropland expansion. Nat Commun 6:8946. https://doi.org/10.1038/ncomms9946

Migdall S, Klug P, Denis A, Bach H (2012) The additional value of hyperspectral data for Smart Farming. IGARSS2012 Munich, IEEE 2012 International Geoscience and Remote Sensing Symposium Proceedings

Migdall S, Spannraft K, Bach H, Hank T, Frank T, Mauser W, Burgstaller S, Tüller G, Angermaier W (2013) On-Farm Application of Operational Integrated Satellite Services, ESA Living Planet Symposium 2013, Special Publication SP-722, Edinburgh, UK

Verhoef W, Bach H (2007) Coupled soil - leaf - canopy and atmosphere radiative transfer modeling to simulate hyperspectral multi - angular surface reflectance and TOA radiance data. Remote Sens Environ 109(2):166–182

3

The Digital Transformation of Education

Ravi Kapur, Val Byfield, Fabio Del Frate, Mark Higgins, and Sheila Jagannathan

Abstract For society to benefit fully from its investment in Earth Observation, the data must be accessible and familiar to a global community of users who have the skills, knowledge and understanding to use the observations appropriately in their work. Future 'Environmental Data Scientists' will need to draw on multiple data and information sources, using data analysis, statistics and models to create knowledge that is communicated effectively to decision-makers in government, industry, and civil society. Networks, cloud computing and visualization will become increasingly important as citizen scientists, data journalists and politicians increasingly use Earth observation products to give their arguments and decisions scientific credibility.

The overarching aim of Earth Observation education must therefore be to support life-long learning, allowing users at all levels to remain up-to-date with EO technologies and communication mechanisms that are relevant to their individual needs. Current and emerging methodologies for interactive education (such as "MOOCs" and mobile learning), and hands-on engagement with real data (such as through citizen science projects) will be central to outreach, training and formal education in this field. To achieve this, it will be important to engage a wider community of experts from a range of disciplines, and to establish a comprehensive network of educators, technical experts, and content producers. It will also be important to encourage "crowd-sourcing" of new contributions, to help maintain scientific and educational quality. A case study from the World Bank's Open Learning Campus illustrates the opportunities to influence thinking much beyond the environmental data scientist community.

R. Kapur (✉)
Imperative Space, London, UK
e-mail: ravi.kapur@imperativespace.com

V. Byfield
National Oceanography Centre, Southampton, UK

F. Del Frate
University of Tor Vergata, Rome, Italy

M. Higgins
EUMETSAT, Darmstadt, Germany

S. Jagannathan
World Bank Group, Washington, DC, USA

Environmental stewardship has "gone mainstream". In just a few short years, public understanding of the vulnerabilities of our environment has moved from marginal to a central part of policy, public discourse and education. The achievement of this tipping point has been so emphatic that there is now a growing and tangible demand, even a thirst, amongst the general population for greater knowledge about how the Earth system works and how to protect it.

EO has been at the heart of the creation of this societal imperative, from the cultural impact of the first Apollo images of the Earth, to the increasing familiarity of astronaut photography and high resolution video from LEO, distributed through social media, news outlets and mainstream media. But there remains a significant disconnect between this increased awareness of EO data, and a detailed understanding of its applications and what it can truly reveal.

To sustain this interest and increase the depth of public and professional understanding of the data, there is a need for new forms of education and training which can cut through a fast-moving and information-rich world. A new generation of decision-makers, social entrepreneurs, educators, media professionals and active citizens are looking to equip themselves quickly and efficiently with deployable, practical knowledge and skills, and the confidence to bring EO data into their work.

Simultaneously, the scientific research community, along with environmental and humanitarian organizations, is in need of a greater through-flow of skilled and knowledgeable data scientists and remote sensing experts. They require the tools to quickly share new ideas, techniques and practice, and provide a window onto emerging developments in EO across multiple disciplines.

Fortuitously, these needs have converged with key developments in a number of arenas which now make all of this possible. Advances and growth in online education, the "open educational resources" (OER) movement (Atkins et al. 2007), and interactive web technologies have coincided with the emergence of open EO data, easier access to high resolution imagery, and a thriving EO app development scene. This has made it possible to provide accessible but authoritative mass education and training and opportunities to work practically with EO data, to diverse audiences around the world.

There are challenges associated with configuring online learning in a way that accommodates a wide array of prior knowledge. Ensuring that the training provided is accessible enough for newcomers but also credible to those with some expertise requires a careful and detailed "learning design" process and clearly signposted pathways to further in-depth learning. But when this is done well, the outcome can be a powerful combination of professional development and academic learning for some learners, and an inspirational and awareness raising tool for new audiences.

A key initial step is to ensure that the learning design is linked to the desired learning outcomes. The process becomes complicated when the course modules target learners who have very different stocks of prior knowledge and different learning objectives, such as raising awareness, professional development or academic learning. A case study from the World Bank's Open Learning Campus (OLC) illustrating how these challenges could be met is included in this paper.

The synergies between the nature of EO education and training, and the growth of interactive online learning create great opportunities for both arenas. The results can often go beyond intended learning, outcomes, with an often emotional and empowered response from those understanding for the first time what EO data can reveal, which in turn may have knock-on societal benefits.

This has been greatly exemplified in a series of EO "MOOCs" (massive open online courses) which have been instigated by ESA. The first two ESA MOOCs (focused on climate monitoring and sensing in the optical realm respectively), along with a course from EUMETSAT (on ocean monitoring) have had over 30,000 registered participants in just 2 years, and further courses are planned. These courses have focused principally on data applications, but have also provided rudimentary training on the basis and history of satellite EO technology and instrumentation, and have acted as a hub for various data access and visualization tools to enable even novice learners to experiment directly with real data.

ESA's first full-scale MOOC, on "Monitoring Climate from Space", was a 5-week course deployed initially on the FutureLearn platform, consisting of a total of 30 original videos, interactive tests and exercises, and 3–5 h of independent study each week. The course attracted 9000 registrants on its first "run", with a participation rate of over 54% and a course completion rate of 22%, above average for large-scale MOOCs on FutureLearn. Over 80% of participants were classified as "active learners" (i.e. following through multiple steps of the course), and around 30% were classified as "social learners", actively involved in peer-learning and online comments. Additionally, the course achieved exceptional qualitative outcomes, with very high approval ratings for the format, quality and "layered" nature of the content, and very high levels of "emotional engagement" and stated intention to continue with further learning in the subject. Significant numbers of newcomers to the subject expressed a profound new appreciation of Earth science, the evidence for climate change and the detail of the data available through satellite observations. Others with existing knowledge of the policy and environmental contexts expressed a renewed intention to incorporate EO into their professional work and decision-making, and an interest in deepening their knowledge of the data and applications (Figs. 1 and 2) (ESA Monitoring Climate from Space MOOC 2015; FutureLearn 2015).

The mix of backgrounds and professions of the course participants was also revealing, ranging across EO and climate scientists, satellite engineers, climate

Fig. 1 Stills from ESA Monitoring Climate from Space MOOC videos. Source: Imperative Space/ESA

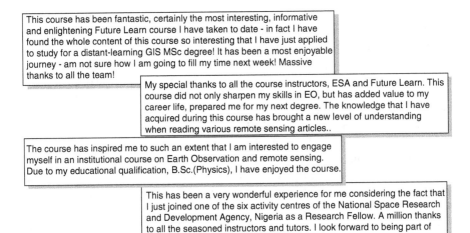

This course has been fantastic, certainly the most interesting, informative and enlightening Future Learn course I have taken to date - in fact I have found the whole content of this course so interesting that I have just applied to study for a distant-learning GIS MSc degree! It has been a most enjoyable journey - am not sure how I am going to fill my time next week! Massive thanks to all the team!

My special thanks to all the course instructors, ESA and Future Learn. This course did not only sharpen my skills in EO, but has added value to my career life, prepared me for my next degree. The knowledge that I have acquired during this course has brought a new level of understanding when reading various remote sensing articles..

The course has inspired me to such an extent that I am interested to engage myself in an institutional course on Earth Observation and remote sensing. Due to my educational qualification, B.Sc.(Physics), I have enjoyed the course.

This has been a very wonderful experience for me considering the fact that I just joined one of the six activity centres of the National Space Research and Development Agency, Nigeria as a Research Fellow. A million thanks to all the seasoned instructors and tutors. I look forward to being part of another course like this from the ESA.

Fig. 2 Example end-of-course learner comments from ESA Monitoring Climate from Space MOOC. Source: Imperative Space/ESA

diplomats, policy-makers, senior decision-makers, school teachers, students (high school, undergraduate, postgraduate and home-educated), campaigners, media professionals, science communicators, retired individuals and casual learners. This demonstrated the potential of MOOCs and OER to address all levels of need in EO education and training, which can be broadly grouped as follows:

1. **Students in higher education** (including those on courses or in research in "adjacent" fields).
2. **Professional development and knowledge transfer** (for existing data scientists, and other professionals working with EO).
3. **School-based education** (especially high school, where links to the curriculum are essential—Early EO education can be seen as "capacity building" for EO skills, two generations hence).
4. **Awareness-raising** and orientation (for newcomers to the field, including decision and policy makers, professional communicators and educators, and the wider public).

All of these areas have a common set of challenges to which new forms of education and training are applicable, but they also have distinct problems which require specific considerations, as outlined below.

In general, EO data is inherently powerful and compelling once understood, but for many potential new audiences it feels inaccessible and highly technical. There are perceived barriers to entry which stem from the technical terminology used, the diversity of data types and the myriad of data portals. It is important to "personalise" the concepts of EO for new audiences and to make the content feel less "remote". "Emotional engagement" and the personal impact of EO imagery and data play an important part in establishing an "embedded" understanding of EO, and in creating a sense of responsibility for the issues the data reveal. In EO education, this level of engagement can be achieved through clear learning design and explanation, but also through high "production values" in the quality of apps, animation, video, infographics, data visualization and "story-telling".

For new and general audiences, OER-based approaches to EO education (especially those which are video-led such as MOOCs) show greatest added value when they serve to "demystify" the subject and increase the visibility of the data and its applications, (i.e. actually "showing" learners how EO technology and data applications work, what the data looks like, and how it can have impact, in science and "on the ground" through real-world case studies).

In the case of higher education, traditional modes of delivery are not well geared to the demands and pace of EO education and training, and radical changes online have already begun to occur which should be embraced by the EO arena. The basic model of higher education existing today was created in the eleventh century, operates on a nineteenth-century calendar and model, and yet is still expected to prepare students for life in the twenty-first century. Technology is now widely changing the game in higher education by leveraging the potential of "social learning". Rapid developments in information and communication technologies (ICT), and in particular the Internet (web 2.0), have enabled *consumers* of content to become *producers* of content, and share this content through social networking sites, blogs, wikis and virtual communities. These technology developments have enabled the emergence of a more participatory approach for open online distance education, especially in the form of MOOCs, and the use of more innovative learning materials (such as videos, audio recordings, interactive tests and collaborative exercises). A popular and early example of the MOOC was MIT's Open Course Ware initiative, which today provides open access to undergraduate and graduate-level materials and modules from more than 1700 courses. Other initiatives, such as the Open University, iTunesU and the Khan Academy, are also key precursors of Open Educational Resources (OER) (Mehaffy 2012; Diaz et al. 2013).

The term "MOOC" was first coined by Dave Cormier in 2008 for a course on "Connectivism and Connective Knowledge". The idea however took off in 2011 when Sebastian Thrun and Peter Norvig, at Stanford University, decided to offer the course "Introduction to Artificial Intelligence" free to anybody in the world. They sent out a single e-mail, and 160,000 people from 190 countries signed up. More than 23,000 students completed the course, and by the end of the course, only 30 of the original 200 Stanford students were still going to face-to-face classes. Out of the 23,000-plus course completers, 248 students had a perfect score in the course, and none of them were Stanford students (Fig. 3) (Mehaffy 2012).

MOOCs are widely regarded as being part of a new paradigm (and possibly revolution) in *"democratization of a new type of education serving the needs of twenty-first century students"*, enabling higher education transition towards digital networked learning. The success of the MIT project led to the creation of Udacity in 2012, and inspired many universities to start their own OER and MOOC initiatives, such as EdX, Coursera, and FutureLearn, to deliver free educational material to millions of people across the world. While some await further evidence for the long term impact of MOOCs, it is apparent that the OER movement is well aligned with the learning expectations of the new generation of higher education students, and

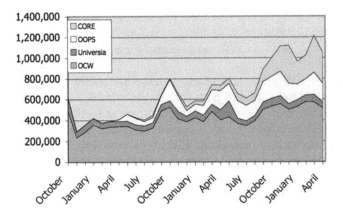

Fig. 3 Monthly visits to MIT OpenCourseWare site, October 2003 to April 2006. Figure above includes OCW translated sites: Universia (Spanish/Portuguese); Open Resources for Education (CORE); Opensource Opencourseware Prototype System (OOPS—Taiwan) (Smith & Casserly 2006)

is particularly well attuned to fast-moving, visual and complex subjects such as EO (Mehaffy 2012).

The ESA MOOCs have in some respects built on the work of previous online EO education projects supported by ESA such as the Learn EO! initiative and the Bilko software. What is different about the MOOCs however is that rather than providing a set of resources tailored for particular target audiences, they are instead designed to offer a spectrum of depth-of-learning for wide ranging audiences. The conundrum for the learning designers has been in finding a way to create a broad offering, whilst responding to the increasing expectation of online learners for a personalized experience. In the ESA courses, a sense of personalization comes in two forms: firstly the "social learning" aspects (including online discussions with other learners and feedback from expert educators); and secondly the range of options available for self-directed learning beyond the core materials of the course.

Others have approached the challenge of personalization of MOOCs in alternative ways. For example, the World Bank Group, which has invested heavily in MOOCs and online learning to help support sustainable development goals, uses online courses as part of a "blended" education and capacity building approach. Their principal target audience is "mid-career" practitioners involved in development, and their aim has been to create a flexible, collaborative learning approach.

This has taken the form of the "Open Learning Campus" (OLC), which includes a combination of full-scale MOOCs, "bitesize" learning resources, short "e-courses" and some face-to-face training (both online and in centres). This has led to a shift in the capacity building training and education delivered by the World Bank online from 5% to 95% over 4 years.

The OLC has so far made available six MOOCs, 500 short courses and 5000 bitesize resources, and their own analysis has concluded that this has enabled 350,000 target learners to access structured education and training, equivalent to around 10,000 face to face training courses delivered in a conventional way.

The flexible array of learning tools provided by the OLC now also includes resources tailored for on-the-move and "just-in-time" mobile learning. They are also pioneering the use of AI to further personalize the learning experience based on prior courses and other preferences.

Over the last 10 years, substantial evidence has started to emerge about the impact of online and video-led professional development training and knowledge transfer in a range of sectors. It has been shown to be particularly effective in sectors where there is a need to break down professional "silos" and enable a cross-disciplinary approach to innovation. This could be said to be very applicable to the training and professional development needs in EO. MOOCs and other forms of structured OER provide an ideal way to quickly unlock critical information and impart know-how for both professional and casual learners.

Professor Richard Elmore from the Harvard Graduate School has described learning as "the conversion of information into knowledge". In fast-innovating fields such as EO, professionals are often operating at the very limit of their knowledge, using the latest innovation they are aware of. To help them incorporate new knowledge into their work, they often need to be literally "shown" latest information and how innovations are being applied elsewhere (Elmore 2015a, b). In the EO arena, the emerging big data revolution will present a challenge for organizing the vast amounts of new data into accessible information, but it will also be a corresponding challenge for EO education to convert that information into tangible, understandable and applicable knowledge.

In the case of knowledge transfer and general awareness raising of EO in other arenas, much can also be learned from the use of OER and video-led training in other areas of science. For example, the UK research funding body EPSRC (Engineering and Physical Sciences Research Council) have, like other science bodies, used video extensively to showcase the capabilities and facilities of their grant recipients to industry, in order to promote and foster new research collaborations. For instance, in 2010/11 EPSRC funded a major UK university, the University of Sheffield, to showcase its large-scale and interdisciplinary research activities, expertise and infrastructure through a series of over 40 short documentary films, accompanied by other forms of text-based information. Watched in their entirety, the videos (presented in six thematic areas relating to major global challenges), represent a form of ad hoc, on-demand training course. In practice, the videos were viewed by their target audience over 150,000 times, were shown at numerous professional and public events and had additional public engagement impacts. The project overall is believed to have led to many significant investments and funded research partnerships, and has been emulated with follow-up video projects of this kind (Research at Sheffield 2010). MOOCs and similar video resources could greatly help to raise the visibility of EO data and applications in this way, amongst policy-makers, senior business executives and funders, effectively training them as "casual" learners.

MOOCs and OER also have a role to play in bringing EO education into high school-based learning, but the approach to how these tools are used has to be modified to take into account the "time poverty" experienced by many teachers, and the need to align resources to the curriculum. School-based education is generally still fairly formalized, and ease of access, in-depth curriculum links and teacher guidance are essential requirements for ensuring take-up in schools. On the other hand, there is often a willingness to embrace technology and create skills-based and personalized learning. EO provides inspirational and compelling learning opportunities in the science, geography and "citizenship" curricula. Educational research has also established that giving students in schools the opportunity to engage directly with real scientific data, (and even contribute to real scientific research), has an extremely positive effect on student engagement, familiarity with key subject matter, follow-on educational interests, and attainment levels. Engaging with real scientific data also helps students to orientate themselves geographically, and provides them with a better understanding of physical aspects of the world around them, such as a sense of scale and rates of change of natural processes. Much of the subject matter in traditional science curricula relates to phenomena which are either too large or too small to be observed directly, so creating a sense of scale and enabling visual understanding of scientific concepts is key to in-depth learning. EO data has the potential to play a significant role in this (Kapur et al. 2007; Facer 2007). At higher levels, the concept can evolve and be scaled up to the Fablab experience, as originally born at MIT's Center for Bits & Atoms Fab Lab Program. In this case, as in the "Fabspace 2.0" project (2016) the main objective is to set up and operate at a university level a free access place and service where students, but also researchers and external people, can make use of a data platform and, under the open innovation umbrella, design and test their own geoinformation applications using EO images.

EO-related citizen science also offers substantial and wide-ranging opportunities in EO education, enabling teachers to explore EO data with their students in a structured manner, and embed it into curriculum delivery and formal teaching. Teachers and students generally relish the opportunity to engage directly with real-world data. It creates depth of learning and insights that are not possible otherwise, and maximises the prospect of students feeling they can make a contribution to science. And there are opportunities for EO education to emulate the successful open research collaborations taking place in other areas of science education. For example, in astronomy a "crowd-sourcing" approach has been established to enable student access to real science infrastructure, such as through the National Schools Observatory project in the UK, and the Faulkes Telescope initiative (Faulkes Telescope Project 2008; National Schools Observatory Project n.d.). Elsewhere, other projects use supercomputer simulation models, or enable school students to engage personally in research side by side with researchers, such as the LUCID project and other initiatives at the Langton Star Centre in the UK (LUCID Project/Langton Star Centre 2014). In lower age groups, EO education is well-placed to capitalize on emerging research into cloud-based education methodologies, such as the "Self-Organised Learning Environments" (or "SOLEs") being pioneered by Professor Sugata Mitra (Dolan et al. 2013).

New variations on the MOOC concept are already emerging which could further widen the potential uses of these approaches in EO education and training. For example, the MOOC tools and format can be delivered to more limited numbers rather than being "massive" (so-called "OOCs"), or distributed to large but private audiences rather than being fully "open" ("MOCs"). Several universities have experimented with the concept of "xMOOCs", where tailored courses are created from a centralized pool of OER and video content, and disseminated through a range of pathways rather than a single platform. Such an approach could be highly applicable to EO education, where content needs to be drawn from multiple agencies and research groups and courses need to be regularly updated. Others are creating courses where content is only accessed "on-demand", with learners encouraged to "re-mix" or "mash-up" the content and share it as they need. What remains essential however, is the need for consistency and clarity of learning design, ensuring ease of access to the content, and supporting opportunities for collaborative, social learning (Diaz et al. 2013; Brown & Adler 2012).

While MOOCs can offer a rich, open and flexible learning experience, they are in effect a new "instructional genre" (Mitchell et al. 2015), and we must use new metrics for assessing their impact and modify our expectations in evaluating success. To many learners using these new tools, "success metrics" and "completion" may not have the same meaning as in traditional education. In the case of EO especially, many learners will be happy to simply access the information they are most interested in, be exposed to new ideas and discover new applications, rather than seek formal accreditation or qualifications (Liyanagunawardena et al. 2014).

Techniques such as MOOCs also enable a two-way flow of insights, helping to inform trainers of what a particular group of learners knows and does not know about EO and the environment, and benefitting the scientific community through the fresh questions and challenges posed by learners. MOOCs include stimulus questions and discussions, and the participants are invited to discuss topics raised in the videos in whatever way they wish. For example, in the EUMETSAT Monitoring the Oceans from Space MOOC, many questions were asked about how the quality of the satellite data is assessed. The participants were using their own language to articulate questions of calibration and validation. These questions provide an indication of how the "story" of creating and managing data sets in science could be told in new ways. The participants seemed to have a real appreciation for the contribution of data validation to the scientific disciplines that rely on the data. As an important side benefit to the main aims of the course, the MOOC showed learners how this key part of the science works by teaching them about specific areas of the science itself.

MOOCs can also be effective in placing the learner at the heart of the science, enabling them to better appreciate the implications of EO data and applications for themselves and their communities. The ESA and EUMETSAT MOOCs have all been designed to highlight connections between the satellite observations, the resulting science and applications, and the impact and opportunities for the individual citizen. Providing tools for participants to create and share examples of their findings is an important element in the teaching process for these MOOCs. This mirrors the process for professionals involved in EO data provision and the efforts to improve the reach of EO data. By encouraging learners to find, obtain and share data with other learners and to comment on what they have learned and what has surprised them, they are more likely to understand why it is not enough to simply provide high quality data, in a stable format and on an easily accessible platform. This practical, interactive, peer learning aspect of the MOOC experience deepens their familiarity with the data and the need to provide tools to visualize, manipulate and communicate about the data for use by wider audiences. It also introduces them to the different ways in which different scientific communities engage with the data (e.g. in meteorology vs. oceanography), and the need for interdisciplinary approaches. In turn, observing how the learners engage with these challenges can provide further insights to inform the development of new interdisciplinary teaching and training.

Case Study: How the World Bank Group's Open Learning Campus is Partnering with Earth Observation Satellite Data to Enrich the Learning Experience

The World Bank Group is a key actor in addressing complex development challenges such as sustainable development, climate change and environmental problems. In this role the organization facilitates training, education and outreach. So too do Earth observation capacity building programs, wherein learners work to understand how sharing and visualizing geospatial data can lead to problem-solving and policy planning for development solutions. The overlap of the need for more real-world learning experiences with context-specific data has led to a partnership between the WBG and Earth observation community that is helping to transform development learning.

Learning as an Accelerator to Achieve Development Goals

The WBG has invested significantly in digital and blended learning, not only relevant to the Earth observation community, but also to a much wider global development community, including policy makers, practitioners, academics, NGOs and the general public. The WBG offers overall online education through its virtual platform, the Open Learning Campus (OLC), through which staff can refresh knowledge on a continuous basis, bridge knowledge gaps jointly with clients and co-create solutions to complex development issues, such as climate change disasters and managing epidemics.

Several drivers led the WBG to invest in the OLC as a go-to destination for development learning. Principally, it is to share this global institution's vast repository of knowledge with the wider development community and thereby transform it into a "Solutions Bank". Internally, such a platform allows for continuous learning on technical, operational and leadership topics to staff, enabling them to remain cutting-edge in their respective fields while also acquiring new competencies that improve their overall effectiveness in the workplace. It also enables uniform career development opportunities for staff, many of whom are in geographically dispersed time zones. Externally, Bank client countries and partners are interested in learning and co-creating solutions, leveraging the repository of development knowledge—especially tacit knowledge—available in the institution.

At the same time several hugely and positively disruptive changes are taking place in the way education and learning is delivered. With close to six billion mobile technology connections projected by 2020 and with about 50% of the world using the Internet[1], the world is increasingly connected and adept at making use of that connectivity for not only entertainment and communication but also for bringing learning tools available any time and at any place. A lot of content is found online through publicly available open educational resources (OER) like Khan Academy's flipped classroom, information downloads possible from Wikipedia, or YouTube. Vast networks of peers are available to respond to queries through Facebook, Snapchat and WhatsApp. Open learning through MOOCs has created learning communities, engaging millions of active learners worldwide. Much of this digital education in developing countries is mobile first, a by-product of the fact that computer ownership is rare for much of the developing world. Digital education is growing 14× faster than traditional higher education[2], suggesting the financial motivation for further change. The increased ability to leverage big data has greatly enhanced learning analytics to understand the contexts in which learning takes place, so that online learning platforms customize learning to meet the specific development needs of learners.

The OLC is converting more and more of WBG's knowledge products and flagship reports (such as the World Development Report and country and regional technical reports) into easily accessible learning products. It is also unlocking tacit knowledge available with many of its staff, clients and partners and packaging this information into absorbable content. Rapid advances in pedagogy and technology have made these innovations possible.

Progress of the OLC so Far

The WBG has historically had a culture of face-to-face learning that is expensive and limited to small numbers of participants. It wanted to move learning away from being an "event" to becoming deeply embedded in its work. As the WBG was shifting to become more knowledge-enabled, it explored how best to deliver knowledge in the form of learning at scale and at significantly lower unit cost. The first foray into e-learning was the establishment of the e-Institute in 2011, which delivered a proof of concept for what later became the OLC. The e-Institute allowed WBG to test client and staff responsiveness and to demonstrate e-learning as a viable

---

[1] http://www.itu.int/en/ITU-D/Statistics/Pages/stat/default.aspx.

[2] https://www.knewton.com/infographics/the-state-of-digital-education-infographic/.

tool to enhance the knowledge and skills of development practitioners who were geographically dispersed.

Encouraged by the results of the e-Institute, WBG launched the OLC in January 2016 with the objective of becoming the preferred destination for development learning. Since the launch, more than two million learners from 190 countries have visited the OLC, and approximately 6000 digital learning activities have been designed and curated, gradually starting to realize the vision of the OLC. In the years to come, the OLC will continue to convene, connect, contextualize and co-create learning for and with our clients and development partners, making the OLC not just a destination for just-in-time content but perhaps even more importantly, a destination for problem-solving conversations and communities.

Flexible Pathways to Learning

The OLC virtual platform leverages advances in pedagogy and technology to provide distinctive formats to match the evolving needs of learners through structured courses, bite-sized lessons and vibrant communities of practice. The OLC harnesses educational technology, mobile capabilities, cloud-computing and innovations in e-learning to produce a world-class learning ecosystem through three "schools" of learning, with robust content, exciting delivery methods and access from anywhere.

The first school, **WB Talks**, enables the learner to explore nuggets of knowledge through talks, podcasts, videos and games. These bite-sized learning opportunities are equally useful for the busy professional and the inquisitive researcher or student.

- Gender and Agriculture (2015)
- Cooperation on the Nile—Bringing Down the Glass Wall (2016)
- Towards Green Growth: Learning from Republic of Korea's Experience (2016)
- Promoting Green Competitiveness (2015)

The second school, **WB Academy**, helps the learner to unpack deep learning related to development challenges and solutions through virtually facilitated or self-directed e-courses, MOOCs and materials from face-to-face courses.

- An Introduction to Land Market Assessment in Complex Urban Settings (2016)
- Designing MRV Systems for Entity-Level Greenhouse Gas Emissions (2016)
- Coffee Price Risk Management (2016)
- Water Footprint—Concept and Application (2016)

The third school, **WB Connect**, enables learners to engage with others through peer and expert learning to find crowd-sourced solutions to development challenges. Using a range of tools, from social media to mobile texts, knowledge exchange can be promoted between WBG staff, clients, partners and global citizens. Examples of such communities include urban floods, Smart Cities and integrated water management and land and house thematic groups.

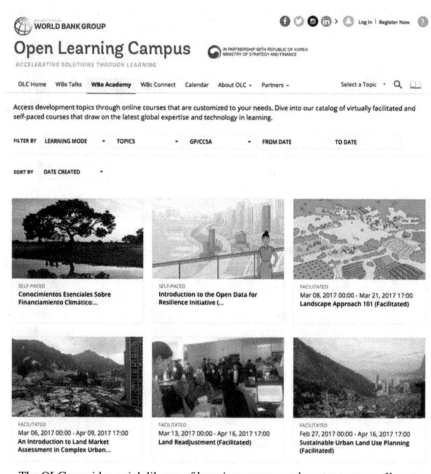

The OLC provides a rich library of learning courses and content across all sectors and regions that is curated from operational groups within the WBG and includes external global knowledge. The learning focuses on best practices and lessons from failures and is facilitated by global and regional experts on development topics from the WBG and the broader development community. The OLC also facilitates opportunities to learn from other client experiences (described as "South-South").

Increasing Importance of Geospatial Data

Various layers of geospatial data, ranging from continuous feeds from Earth observation satellites to geo-coded information on specific locations made available through sensors, are now retrievable and used for development policy and planning applications. Adding a geographic or spatial dimension gives the development learner greater authenticity in terms of the learning experience, while data visualization enables learners to experience realistic simulations by contextualizing information.

For example, learners are able to track air pollution in congested road arteries by the time of the day, send feedback through their mobile phones, and join discussions with traffic planning experts on the best ways to mitigate congestion and air pollution in the streets they travel in. Public participation in planning new facilities in a city, such as a park or shopping mall, could benefit from citizens who are able to visualize spatial outcomes of different proposals under consideration.

Examples of How the OLC is Incorporating Spatial Data

MOOCs are powerful vehicles to distil and share knowledge among various stakeholders, ranging from policy makers to the general public. The OLC offers the MOOC, From Climate Science to Action, thrice in English, twice in Spanish and now under finalization in French. The course guides learners through the latest research that forms the basis for predictions around the effects of climate change on their lives. It explores the projected effects of climate change across sectors, with in-depth analysis of each world region through a series of region-specific modules. Understanding how levels of carbon emissions today will affect the world tomorrow galvanizes public support for political action.

Themes from two European Space Agency (ESA) MOOCs have been incorporated into the WBG climate change MOOC. Videos from the ESA MOOC, "Monitoring Climate Change from Space," have been used to explain Earth observation concepts, how climate data from satellites are valuable inputs in policy formulation and the importance of Earth observation megatrends in wind patterns, ocean currents, etc. for policy applications and promoting global dialogues. Videos from the ESA MOOC, "Earth Observation: The Optical View," have been used to help learn about causality processes in climate change and appreciate the impacts in terms of changes in land use, forest cover, etc.

A second vehicle the OLC offers is facilitated e-learning courses. A series of facilitated e-learning courses in disaster risk management have considerable inputs from Earth observation satellites. An introductory course familiarizes learners with contemporary concepts and practices in disaster management and explains how planning approaches have moved from reactive to pro-active modes. Another course on safe and resilient cities provides urban planners with tools on vulnerability assessments, policies and programs, stock-taking and gaps analysis and financial access. The discussion also focuses on decision-making in uncertainty to identify options and establish priorities, making the city more resilient to climate change impacts and natural disasters. The seven modules of the "Sustainable Urban Land Use Planning" course give the learner a thorough grounding in best practices in urban land use planning and how to sequence regulations and investments to mitigate disaster risks.

A third vehicle is bite-sized learning modules that provide just-in-time learning that complements the above two categories in bringing about meaningful change across all levels—from grassroots action to national policy making to international cooperation. A good example is the Spatial Agent Tutorial, which teaches how to curate data from hundreds of data sets to create on-demand visualization on wide-ranging topics of interest to the Earth observation satellite community, ranging from transport and trade to watersheds and climate change. Bite-sized learning on "Bits and Bricks" focuses on cities undergoing profound transformations, and on how the convergence of digital information (bits) and physical environment (bricks) are facilitated through the Internet to develop new solutions. In support of this initiative, the OLC has prepared a series of bite-sized learning videos that benefited from the Bits and Bricks conference organized by the WBG, with MIT and World Economic Forum.

Issues Relevant to This Experience for the Earth Observation Community

Key issues and messages gathered from deploying training and education in the OLC so far include:

- **Transformation in learning**: Learning is an important accelerator to eradicate poverty and boost shared prosperity whenever global knowledge is converted into practical actions through evidence-based and iterative learning. The OLC transforms how its target audiences learn. It integrates innovations in technology and instructional design to provide accessible, quality learning in a resource-efficient way. Emerging immersive learning modalities facilitate this transformation, by accelerating learning adoption and application. Examples include serious games, geospatial data, artificial intelligence and mobile- and cloud-based learning.
- **Role of Earth observation data**: The OLC provides a unique opportunity for the Earth observation community to disseminate their professional insights to the wider development community for cross-sectoral, "anytime and anywhere" learning. Spatial and Earth observation data are becoming increasingly important in several areas of WBG engagement, notably related to climate change, infrastructure service planning and delivery and citizen engagement because the costs of installing sensors are falling dramatically.
- **Digital changes and connectivity**: Global connectivity is increasing, leading to a "Facebook nation" of more globalized, coordinated, economic, social and political actions across borders. As WBG and its partners tackle complex challenges, many staff members and officials need to learn new skills to leverage the massive expansion in data available—not only from Earth observation satellites but also from mobile call records and sensors on the ground and under water. This data will be valuable only when it can be applied to policy formulation and project design.
- **Wider dissemination**: Partnerships are emerging with regional and national institutions and donors (such as the Korea Development Institute, Monterrey Tec,

Chinese Academy of Governance, Indian National Institute of Urban Affairs) to localize courses and scale up in-country. The know-how can be disseminated globally, with the OLC playing a facilitator role.

- **Citizen engagement**: Influential NGO networks are also participating and contributing to the OLC, so that citizens have the opportunity to acquire the necessary knowledge and skills, and burgeoning citizen movements are persuaded to support the right kind of reforms.

Case Study Conclusion

The OLC represents the WBG commitment to harnessing the global revolution in learning to lift people out of poverty and to boost shared prosperity. Adding the geographic or spatial dimension to the OLC's learning offerings is invaluable because it provides greater authenticity to the learning experience, through big data analytics with data visualization. These types of data enable learners to undertake realistic simulations of complex policy choices and will contribute to more effective decision-making.

References

An Introduction to Land Market Assessment in Complex Urban Settings (2016) https://olc.worldbank.org/content/introduction-land-market-assessment-complex-urban-settings-facilitated

Atkins DE, Brown JS, Hammond AL (2007) A review of the open educational resources (OER) movement: achievements, challenges, and new opportunities

Brown JS, Adler RP (2012) Minds on fire: open education, the long tail, and learning 2.0

Coffee Price Risk Management (2016) https://olc.worldbank.org/content/coffee-price-risk-management-self-paced

Cooperation on the Nile—Bringing Down the Glass Wall (2016) https://olc.worldbank.org/content/cooperation-nile-bringing-down-glass-wall

Designing MRV Systems for Entity-Level Greenhouse Gas Emissions (2016) https://olc.worldbank.org/content/designing-mrv-systems-entity-level-greenhouse-gas-emissions-self-paced

Diaz V, Brown M, Pelletier S (2013) Learning and the massive open online course. A report on the ELI focus session. EDUCSAUSE review

Dolan P, Leat D, Mazzoli Smith L, Mitra S, Todd L, Wall K (2013) Self-organised learning environments (SOLEs) in an English school: an example of transformative pedagogy? Online Educ Res J

Elmore RF (2015a) School reform from the inside out - policy, practice, and performance. Harvard Education Press, Cambridge, MA

Elmore RF (2015b) The future is learning, but what about schooling. Inside Higher Ed

ESA Monitoring Climate from Space MOOC (2015) Imperative Space/ESA. https://www.futurelearn.com/courses/climate-from-space

"Fabspace 2.0" project (2016) https://www.fabspace.eu/

Facer K (2007) Principles for consideration in the design of future learning environments. Futurelab

Faulkes Telescope Project (2008) http://www.faulkes-telescope.com

FutureLearn (2015) Monitoring climate from space – course report

Gender and Agriculture (2015) https://olc.worldbank.org/content/gender-and-agriculture

Kapur R, Nieman A et al (2007) Project Faraday – exemplar designs for science. Department for Children Schools and Families, UK Government, London

Liyanagunawardena TR, Parslow P, Williams SA (2014) Dropout – MOOC participants' perspective

LUCID Project/Langton Star Centre (2014) http://www.thelangtonstarcentre.org/lucid/

Mehaffy GLM (2012) Challenge and change. EDUCSAUSE review

Mitchell J, Stevens M, Thille C (2015) What we've learned from MOOCs, Insider Higher Ed/Lytics Lab, Stanford University, Stanford, CA

National Schools Observatory Project (n.d.) http://www.schoolsobservatory.org.uk

Promoting Green Competitiveness (2015) https://olc.worldbank.org/content/lightning-talk-promoting-green-competitiveness

Research at Sheffield (2010) GovEd Communications/EPSRC/University of Sheffield. http://www.researchatsheffield.co.uk

Smith MS, Casserly CM (2006) The promise of open educational resources. Change 38(5):8

Towards Green Growth: Learning from Republic of Korea's Experience (2016) https://olc.worldbank.org/content/towards-green-growth-learning-republic-korea%E2%80%99s-experience

Water Footprint—Concept and Application (2016) https://olc.worldbank.org/content/water-footprint-concept-and-application-self-paced

4

Machine Learning Applications for Earth Observation

David J. Lary, Gebreab K. Zewdie, Xun Liu, Daji Wu, Estelle Levetin,
Rebecca J. Allee, Nabin Malakar, Annette Walker, Hamse Mussa,
Antonio Mannino, and Dirk Aurin

Abstract Machine learning has found many applications in remote sensing. These applications range from retrieval algorithms to bias correction, from code acceleration to detection of disease in crops, from classification of pelagic habitats to rock type classification. As a broad subfield of artificial intelligence, machine learning is concerned with algorithms and techniques that allow computers to "learn" by example. The major focus of machine learning is to extract information from data automatically by computational and statistical methods. Over the last decade there has been considerable progress in developing a machine learning methodology for a variety of Earth Science applications involving trace gases, retrievals, aerosol products, land surface products, vegetation indices, and most recently, ocean applications. In this chapter, we will review some examples of how machine learning has already been useful for remote sensing and some likely future applications.

Introduction

Beyond remote sensing, machine learning has already proved immensely useful in a wide variety of applications in science, business, health care, and engineering. Machine learning allows us to *learn by example*, and to *give our data a voice*. It is particularly useful for those applications for which we do *not* have a complete theory, yet which are of significance. Machine learning is an automated implementation of the scientific method (Domingos, 2015), following the same process of generating, testing, and discarding or refining hypotheses. While a scientist or engineer may spend his entire career coming up with and testing a few hundred hypotheses, a machine-learning system can do the same in a fraction of a second. Machine learning provides an objective set of tools for automating discovery. It is

D.J. Lary (✉) • G.K. Zewdie • X. Liu • D. Wu • E. Levetin • R.J. Allee • N. Malakar
A. Walker • H. Mussa • A. Mannino • D. Aurin
Hanson Center for Space Sciences, The University of Texas at Dallas, 800 West Campbell Road, Richardson, TX 75080, USA
e-mail: david.lary@utdallas.edu; https://davidlary.info

therefore not surprising that machine learning is currently revolutionizing many areas of science, technology, business, and medicine (Lary et al., 2016).

Machine learning is now being routinely used to work with large volumes of data in a variety of formats such as image, video, sensor, health records, etc. Machine learning can be used in understanding this data and creating predictive and classification tools. When machine learning is used for regression, empirical models are built to predict continuous data, facilitating the prediction of future data points, e.g. algorithmic trading and electricity load forecasting. When machine learning is used for classification, empirical models are built to classify the data into different categories, aiding in the more accurately analysis and visualization of the data. Applications of classification include facial recognition, credit scoring, and cancer detection. When machine learning is used for clustering, or unsupervised classification, it aids in finding the natural groupings and patterns in data. Applications of clustering include medical imaging, object recognition, and pattern mining. Object recognition is a process for identifying a specific object in a digital image or video. Object recognition algorithms rely on matching, learning, or pattern recognition algorithms using appearance-based or feature-based techniques. These technologies are being used for applications such as driver-less cars, automated skin cancer detection, etc.

There are now a variety of open source tools that can greatly facilitate the use of machine learning, such as scikit-learn,[1] TensorFlow,[2] Caffe,[3] and Spark Mlib.[4] Common programming environments used for machine learning include R,[5] Python,[6] and Matlab.[7] All of the applications shown in this chapter used matlab.

In this paper we will give an overview of several remote sensing applications of machine learning made over the last two decades and then take a look ahead to some likely future applications.

What Is Machine Learning?

Machine learning is an automated approach to building empirical models from the data *alone*. A key advantage of this is that we make *no* a priori assumptions about the data, its functional form, or probability distributions. It is an empirical approach, so we do not need to provide a theoretical model. However, it also means that for machine learning to provide the best performance we do need a

---

[1] http://scikit-learn.org/stable/.

[2] https://www.tensorflow.org.

[3] http://caffe.berkeleyvision.org.

[4] http://spark.apache.org/mllib/.

[5] https://cran.r-project.org.

[6] https://www.python.org.

[7] https://www.mathworks.com/solutions/machine-learning.html.

comprehensive representative set of examples, that spans as much of the parameter space as possible. This comprehensive set of examples is referred to as the *training data*.

So, for a successful application of machine learning we have *two* key ingredients, both of which are essential, a machine learning algorithm, and a comprehensive training data set. Then, once the training has been performed, we should test its efficacy using an independent validation data set to see how well it performs when presented with data that the algorithm has *not* previously seen, i.e. test its *generalization*. This can be, for example, a randomly selected subset of the training data that was held back and then utilized for independent validation.

It should be noted that with a given machine learning algorithm, the performance can go from poor to outstanding with the provision of a progressively more complete training data set. Machine learning really is learning by example, so it is critical to provide as complete a training data set as possible. At times, this can be a labor-intensive endeavor.

When using machine learning we are typically performing one of three tasks:

1. Multivariate non-linear non-parametric regression.
2. Supervised classification.
3. Unsupervised classification.

Each of these tasks can be achieved by a variety of different algorithms. Some of the commonly used algorithms include Neural Networks (McCulloch and Pitts, 1943; Haykin, 2001, 2007, 1994, 1999; Demuth et al., 2014; Bishop, 1995), Support Vector Machines (Vapnik, 1982, 1995, 2000, 2006; Cortes and Vapnik, 1995), Decision Trees (Safavian and Landgrebe, 1991), and Random Forests (Ho, 1998; Breiman, 1984, 2001).

Let us now turn our attention to some examples.

Some Existing Machine Learning Applications

We will start by looking at several examples of bias correction. Bias identification and correction is of particular importance for every single remote sensing instrument. Bias correction can also prove to be a particularly challenging issue, one which involves multiple factors.

Machine Learning for Bias Correction and Cross Calibration

The ubiquitous issue of inter-instrument biases is an obvious example of where we do *not* have a complete theoretical understanding, and so machine learning can be of particular use.

In many areas of remote sensing we have multiple instruments simultaneously observing the earth on a variety of platforms. Many of these sensors may be providing data on the same parameters, such as the surface vegetation, the composition of the atmosphere or ocean. A ubiquitous issue faced is inter-instrument bias between the contemporaneously observing instruments. This inter-instrument bias can be due to a variety of known reasons that may include different instruments, different observing geometry and orbits, etc., as well as some causes that we do not know.

This is an important issue, as we routinely need to provide data fusion of multiple datasets. Datasets which are inevitably biased relative to each other, sometimes even after the mandatory calibration/validation process. When we are seeking to construct a long-term record spanning many decades this inevitably will often involve a large number of instruments, a matter very relevant for climate variables. In addition, Data Assimilation has become an important part of effectively utilizing remotely sensed data. However, data assimilation is a *Best Linear Unbiased Estimator* (BLUE), and fusing biased data can cause serious issues.

This data fusion typically involves large teams of scientists and engineers. On the one hand, the instrument teams have a keen sense of faithfully reporting the data, as it is, warts and all. They are naturally loath to empirically correct biases; they would like to theoretically understand the cause of the bias and data issues from first principles. However, as the Earth System is so complex, with many interacting processes, and often the instruments are also complex, this is not always possible. Residual data issues can, and usually do, remain. On the other hand, the modelers know that data bias exists, but are very reticent to make changes to data products that they did not collect, so we therefore have *a problem of closure*.

Biases are ubiquitous, not all of them can be explained theoretically. Yet, we typically need to fuse multiple datasets to construct long-term time series and/or improve global coverage. If the biases are not corrected before data fusion we introduce further problems, such as spurious trends, leading to the possibility of unsuitable policy decisions. When data assimilation is involved, any use of biased observations can lead to the sub-optimal use of the observations, non-physical structures in the analysis, biases in the assimilated fields, and extrapolation of biases due to multivariate background constraints. To compound matters further, the instruments whose data we would like to fuse are often not making coincident measurements in time or space. It is imperative to inter-compare observations in their appropriate context and be able to address the pernicious issue of inter-instrument bias. An issue where machine learning has proved to be most useful. Let us now take a look at some examples.

Vegetation Indices

Consistent, long-term vegetation data records are critical for analysis of the impact of global change on terrestrial ecosystems. Continuous observations of terrestrial ecosystems through time are necessary to document changes in magnitude or variability in an ecosystem. Satellite remote sensing has been the primary tool for

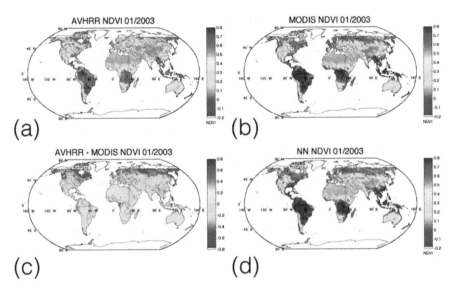

Fig. 1 Using the same color scale, panels (**a**) and (**b**) show the contemporaneous January 2003 NDVI averages for AVHRR and MODIS, respectively. The differences are particularly evident in the higher northern latitudes. Panel (**c**) shows the absolute difference in NDVI between AVHRR and MODIS. Panel (**d**) shows the *estimated* MODIS NDVI for January 2003 using AVHRR data and a neural network. The neural network has been extremely effective in removing the substantial bias in NDVI between AVHRR and MODIS

scientists to measure global trends in vegetation, as the measurements are both global and temporally frequent. To extend measurements through time, multiple sensors with different design and resolution must be used together in the same time series. This presents significant problems as sensor band placement, spectral response, processing, and atmospheric correction of the observations can vary significantly and impact the comparability of the measurements.

Even without differences in atmospheric correction, vegetation index values for the same target recorded under identical conditions will not be directly comparable because input reflectance values differ from sensor to sensor due to differences in sensor design and spectral response of the instrument. This is clearly visible in the example shown in Fig. 1. Using the same color scale, panels (a) and (b) show the contemporaneous January 2003 NDVI averages for AVHRR and MODIS, respectively. The differences are particularly evident in the higher northern latitudes. Panel (c) shows the difference in NDVI between AVHRR and MODIS, there are substantial biases present.

Brown et al. (2008) showed how machine learning, in particular, neural networks, can identify and remove differences in sensor design and variable atmospheric contamination from the AVHRR NDVI record in order to match the range and variance of MODIS NDVI without removing the desired signal representing the underlying vegetation dynamics. This is well illustrated by comparing Fig. 1 panels (b) and (d). Panel (b) shows the actual MODIS NDVI for January 2003. Panel

(d) shows the *estimated* MODIS NDVI for January 2003 using AVHRR data and a neural network, they are almost indistinguishable. The neural network has been extremely effective in removing the substantial bias in NDVI between AVHRR and MODIS.

Neural networks are "data transformers," where the objective is to associate the elements of one set of data to the elements in another. Relationships between the two datasets can be complex and the two datasets may have different statistical distributions. This transformation incorporates additional input data that may account for differences between the two datasets.

Brown et al. (2008) demonstrated the viability of neural networks as a tool to produce a long-term dataset based on AVHRR NDVI that has the data range and statistical distribution of MODIS NDVI. Previous work has shown that the relationship between AVHRR and MODIS NDVI is complex and nonlinear, thus this problem is well suited to neural networks if appropriate inputs can be found. The impact of atmospheric contamination, such as clouds, smoke, pollution, and other aerosols, variations in soil color and exposure through vegetation, and land cover type has a differential effect on AVHRR data as compared to MODIS data. Brown et al. (2008) used overlapping years of observations to train the neural networks.

Remote sensing datasets are the result of a complex interaction between the design of a sensor, the spectral response function, stability in orbit, the processing of the raw data, compositing schemes, and post-processing corrections for various atmospheric effects including clouds and aerosols. The interaction between these various elements is often nonlinear and non-additive, where some elements increase the vegetation signal-to-noise ratio (compositing, for example) and others reduce it (clouds and volcanic aerosols). Thus, although many have used simulated data to explore the relationship between AVHRR and MODIS, these techniques are not directly useful in producing a sensor-independent vegetation dataset that can be used by data users in the near term.

There are substantial differences between the processed vegetation data from AVHRR and MODIS. In order to have a long data record that utilizes all available data back to 1981, we must find practical ways of incorporating the AVHRR data into a continuum of observations that include both MODIS and VIIRS. Brown et al. (2008) showed that the TOMS data record on clouds, ozone, and aerosols can be used to identify and remove sensor-specific atmospheric contaminants that differentially affect the NDVI from AVHRR over MODIS. Other sensor-related effects, particularly those of changing BRDF, viewing angle, illumination, and other effects that are not accounted for here, remain important sources of additional variability. Although this analysis has not produced a dataset with identical properties to MODIS, it has demonstrated that a neural net approach can remove most of the atmospheric-related aspects of the differences between the sensors, and match the mean, standard deviation, and range of the two sensors. A similar technique can be used for the VIIRS sensor.

Let us now look at some other examples related to the remote sensing of atmospheric composition.

Space-Based Measurements of HCl Relevant for Ozone Depletion

The magnitude of atmospheric ozone depletion is closely intertwined with the abundance of atmospheric halogens such as chlorine. The main reservoir for atmospheric chlorine compounds is HCl. The peak in stratospheric HCl was reached in the late 1990s. Between 1998 and 2004 the stratospheric loading of HCl was relatively constant, with some month to month fluctuation; this was followed by a

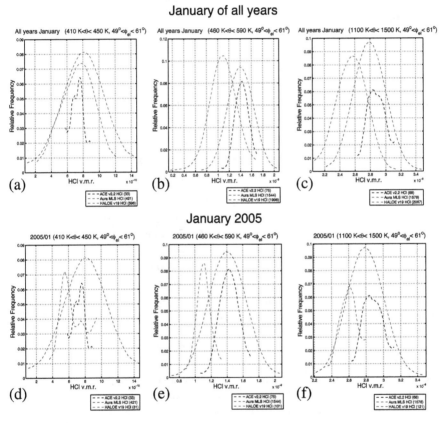

Fig. 2 Example HCl PDFs for the three instruments HALOE, ACE, and Aura MLS. In each case the PDFs are for all observations made by that instrument in a Lagrangian region for three isentropic levels centered on an equivalent latitude of 55°N during all the Januarys that the instrument observed. For ACE the plots include January 2004–2006, for HALOE the plots include 2004–2005, and for MLS the plots include 2005–2006. (**a**) Plot of a PDF for all observations in the range $410\,\mathrm{K} < \theta < 450\,\mathrm{K}$ ($70\,\mathrm{mbar} < P < 110\,\mathrm{mbar}$), $49° < \phi_{el} < 61°$. (**b**) Plot of a PDF for all observations in the range $460\,\mathrm{K} < \theta < 590\,\mathrm{K}$ ($30\,\mathrm{mbar} < P < 60\,\mathrm{mbar}$), $49° < \phi_{el} < 61°$. (**c**) Plot of a PDF for all observations in the range $1100\,\mathrm{K} < \theta < 1500\,\mathrm{K}$ ($2\,\mathrm{mbar} < P < 4\,\mathrm{mbar}$), $49° < \phi_{el} < 61°$. Panels (**d**)–(**f**) are analogous to panels (**a**)–(**c**) for the observations made only during January 2005. The number of observations used to form each PDF is shown in parentheses in the legend

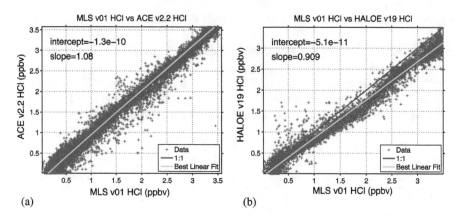

Fig. 3 Panels (**a**) and (**b**) show scatter plots of all contemporaneous observations of HCl made by HALOE, ACE, and Aura MLS. Each point plotted is the median value of a PDF of observations made for a Lagrangian region over the period of a month. It can be seen that to adequately using the slopes to describe the differences does not do justice to the differences. For example, panel (**a**) shows a much better agreement than panel (**b**), but the slopes themselves do not reflect this. Panel (**b**) shows an offset of order 9% near 2 ppbv, whereas panel (**a**) shows maybe 1 or 2% for values near 2 ppbv. The mean absolute value of the differences seems a good indicator of the fits

more pronounced decrease in HCl since 2004. As can be seen in Figs. 2 and 3, we can use probability distribution functions (PDFs) and scatter diagrams for validation and bias characterization of Aura Microwave Limb Sounder (MLS) HCl retrievals. Both these methods allow us to use large statistical samples and do not require correlative measurements to be collocated in space and time.

We can take the difference between the medians of the PDFs as a measure of the inter-instrument bias. This bias is really only significant if it is larger than the atmospheric variability in the Lagrangian region we are considering (i.e., the width of the PDF).

We compared the PDFs for all overlapping Lagrangian regions for a given month. However, we can use a single scatter diagram to compare all the overlaps globally for all the months observed by each pair of instruments (Fig. 3). Such a scatter diagram has the advantage of a *huge sample size*, it encompasses the entire period that a pair of instruments were making contemporaneous observations. The scatter diagram is intended as a big picture summary for all contemporaneous observations made globally. Figure 4 shows two scatter diagrams for all the contemporaneous observations of HCl made globally by two pairs of instruments. In Fig. 4a we compare ACE and Aura MLS which were making contemporaneous observations between September 2004 and the present. In Fig. 4b we compare HALOE and Aura MLS which were making contemporaneous observations between September 2004 and November 2005.

In the ideal case where we have perfect agreement between two instruments, the slope of the scatter diagram would be 1 and the intercept would be 0. In the case of ACE and Aura, we see there is a slope of 1.08, and for HALOE Aura there is a

Training   Independent Validation

Recalibrating Aura MLS HCl to agree with ACE v2.2 HCl

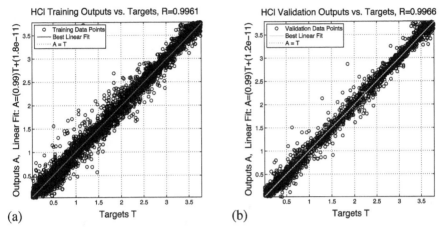

(a)   (b)

Recalibrating HALOE v19 HCl to agree with ACE v2.2 HCl

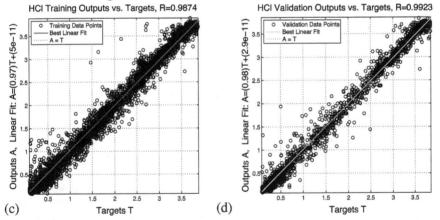

(c)   (d)

Fig. 4 (left) Result of training a neural network to learn the interinstrument biases. (right) An independent validation of this training using a randomly chosen, totally independent, data sample not used in training the neural network. In each case, the x axis shows the actual ACE v2.2 HCl (the target). (**a**) and (**b**) The y axis is the neural network estimate of ACE v2.2 HCl based on Aura MLS v01 HCl. Panel (**a**) is the result using the training data, and Fig. 5b is the result of the independent validation. Panels (**c**) and (**d**) The y axis is the neural network estimate of ACE v2.2 HCl based on HALOE v19 HCl. Figure 5c is the result using the training data, and panel (**d**) is the result of the independent validation. In each case, this is a global training for all contemporaneous observations between each pair of instruments. The training points are the median values of a PDF of observations made during a given month for a given equivalent PV latitude—potential temperature bin. The width of the cloud of points in each of these scatter diagrams is a good measure of the uncertainty associated with the neural network fit

slope of 0.91 (Fig. 3). It can be seen that solely using the slopes does not do justice to the differences. For example, Fig. 3a shows a much better agreement than Fig. 3b, but the slopes themselves do not reflect this. Figure 3b shows an offset of order 9% near 2 ppbv, whereas Fig. 3a shows maybe 1 or 2% for values near 2 ppbv. The mean absolute value of the differences seems a good indicator of the fits. We also note that in the case of Aura MLS and HALOE, the scatter diagrams do not have a constant slope over the entire range of HCl values, several "wiggles" are present. This means that the inter-instrument biases are *spatially and temporally dependent*. Neural networks are multi-variate, non-parametric, "learning" algorithms that are ideally suited to learning, and correcting for, such inter-instrument biases.

We have used a neural network with three inputs and one output. The inputs are equivalent PV latitude, potential temperature, and HCl from instrument A. The output is HCl from instrument B. Potential temperature and equivalent latitude are used because they are good markers of the large-scale flow pattern. When we do the training we randomly split our training data set into three portions of 80%, 10%, and 10%. The 80% is used to train the neural network weights. This training is iterative and on each iteration we evaluate the current RMS error of the neural network. The RMS error is calculated by using the second 10% of the data that was not used in the training. We use the RMS error and the way it changes with training iteration (epoch) to determine the convergence of our training. When the training is complete, we use the final 10% as a validation data set. This 10% of the data was randomly chosen and not used in either the training or RMS evaluation. We only use the neural network if the validation scatter diagram, which plots the actual data from validation portion against the neural network estimate, yields a straight line graph with a slope of 1. This is a stringent and independent validation. The validation is global as the data was randomly selected over all temporal and spatial data points available. Several training strategies were examined, the one described included the most species over the longest time period. The neural network algorithm used was a feed-forward back-propagation network with 20 hidden nodes. The training was done by the Levenberg-Marquardt back-propagation algorithm.

Figure 4 shows the results of such a neural network training to learn inter-instrument biases between ACE v2.2, Aura MLS v1 and HALOE v19 HCl. Panels (b) and (d) show an independent validation of the training using a randomly chosen, totally independent, data sample not used in training the neural network. In each case the x axis shows the actual ACE v2.2 HCl (the target). In panels (a) and (b) the y axis is the neural network estimate of ACE v2.2 HCl based on Aura MLS v01 HCl. Panel (a) shows the results using the training data, panel (b) shows the results of the independent validation. In panels (c) and (d) the y axis is the neural network estimate of ACE v2.2 HCl based on HALOE v19 HCl. Panel (c) shows the results using the training data, and panel (d) shows the results of the independent validation. The mapping has removed the bias between the measurements and has also straightened out the "wiggles" seen in Fig. 3.

The bias between the Halogen Occultation Experiment (HALOE) and Aura MLS is greatest above the 525 K (21 km) isentropic surface. The global average mean bias between Aura and the Atmospheric Chemistry Experiment (ACE) for January 2005 was 2% and between Aura MLS and HALOE was 14%. The widths of the PDFs are a measure of the spatial variability and measurement precision. The Aura MLS HCl PDFs are consistently wider than those for ACE and HALOE, this reflects the retrieval uncertainties. The median observation uncertainty for Aura MLS v1.51 HCl is 12%, and the median ACE v2.2 uncertainty is 8%. We also connect Aura MLS HCl with the heritage of HALOE HCl by using neural networks to learn the inter-instrument biases and provide a seamless HCl record from the launch of the Upper Atmosphere Research Satellite (UARS) in 1991 to the present (Lary and Aulov, 2008).

HCl and Cl_y Time Series

Knowledge of the distribution of inorganic chlorine Cl_y in the stratosphere is needed to attribute changes in stratospheric ozone to changes in halogens, and to assess the realism of chemistry-climate models. However, there are limited direct observations of Cl_y. Simultaneous measurements of the major inorganic chlorine species are rare. In the upper stratosphere, Cl_y can be inferred from HCl alone.

Now that we have completely characterized the inter-instrument biases and been able to correct for them we can connect Aura MLS HCl observations to the heritage of HALOE (Lary et al., 2007). This allows us to produce an HCl time series from the launch of UARS in 1991 up to the present. Figure 5 shows HCl time series for six different locations with HCl observations from HALOE, ATMOS, ACE, MkIV and Aura MLS.

The HCl re-calibrations have been used (Fig. 6) to form a long Cl_y time series and associated uncertainty estimate (typically 0.4 ppbv at 800 K). The uncertainty in the Cl_y estimate is primarily due to the discrepancy between the different observations of HCl, i.e., the HALOE, Aura MLS, and ACE inter-instrument biases.

A consistent time series of stratospheric Cl_y from 1991 (Fig. 6) has been formed using available space-borne observations (Lary et al., 2007). Here we used neural networks to inter-calibrate HCl measurements from different instruments, and to estimate Cl_y from observations of HCl and CH_4. These estimates of Cl_y peaked in the late 1990s and have begun to decline as expected from tropospheric measurements of source gases and troposphere to stratosphere transport times. Furthermore, the estimates of Cl_y are consistent with calculations based on tracer fractional releases and age of air. The Cl_y time-series formed here is an important benchmark for models being used to simulate the recovery of the ozone hole. Although there is uncertainty in the estimates of Cl_y, primarily due to biases in HCl measurements, this uncertainty is small compared with the range of model predictions shown in the 2006 WMO report (Lary et al., 2007). This work was viewed as ground breaking and received three awards as a JCET science highlight, a NASA Aura Mission Science highlight, and as a NASA GSFC Atmospheric Chemistry and Dynamics Branch selected publication.

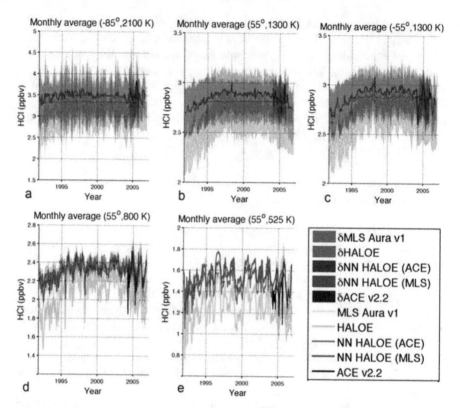

Fig. 5 Selected HCl time series from the launch of UARS to 2007 with HCl observations from HALOE, ATMOS, ACE, and Aura. (**a**) For 2100 K (50 km) at 85°S, (**b**) and (**c**) for 1300 K (41 km) at 55°N and 55°S, and (**d**) and (**e**) for 55°N at 800 K (30 km) and 525 K (21 km). In each case the green line and shading is for the original HALOE v19 data, and the red line and shading is for HALOE data remapped with a neural network to agree with Aura MLS v1 HCl. The black line is the ACE v2.2 data with the grey shading representing the associated uncertainty. The d in the legend which labels the shading refers to the total uncertainty (observational, representativeness, and where relevant, neural network adjustment). The remapping of HALOE generally brings the HALOE data into better agreement with the ATMOS (squares) data

Bias Correction of MODIS Aerosol Optical Depth

Aerosol and cloud radiative effects remain some of the largest uncertainties in our understanding of climate change. Over the past two decades, observations and retrievals of aerosol characteristics have been conducted from space-based sensors, from airborne instruments, and from ground-based samplers and radiometers. Much effort has been directed at these data sets to collocate observations and retrievals and to compare results. Ideally, when two instruments measure the same aerosol characteristic at the same time, the results should agree within well-understood measurement uncertainties. When inter-instrument biases exist, we would like to explain them theoretically from first principles. One example of this task is the

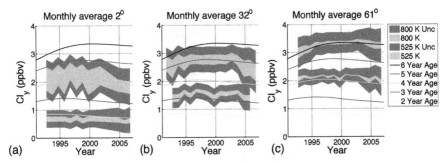

Fig. 6 (**a–c**) October Cl$_y$ time-series for the 525 K isentropic surface ($\approx$20 km) and the 800 K isentropic surface ($\approx$30 km). In each case the dark shaded range represents the total uncertainty in our estimate of Cl$_y$. This total uncertainty includes the observational uncertainty, the representativeness uncertainty (the variability over the analysis grid cell), the inter-instrument bias in HCl, the uncertainty associated with the neural network inter-instrument correction, and the uncertainty associated with the neural network inference of Cl$_y$ from HCl and CH$_4$. The inner light shading depicts the uncertainty on Cly due to the inter-instrument bias in HCl alone. The upper limit of the light shaded range corresponds to the estimate of Cl$_y$ based on all the HCl observations calibrated by a neural network to agree with ACE v2.2 HCl. The lower limit of the light shaded range corresponds to the estimate of Cl$_y$ based on all the HCl observations calibrated to agree with HALOE v19 HCl. Overlaid are lines showing the Cl$_y$ based on age of air calculations. To minimize variations due to differing data, coverage months with less than 100 observations of HCl in the equivalent latitude bin were left out of the time-series

comparison between the aerosol optical depth (AOD) retrieved by the MODerate resolution Imaging Spectroradiometer (MODIS) and the AOD measured by the Aerosol Robotic Network (AERONET). While progress has been made in understanding the biases between these two data sets, we still have an imperfect understanding of the root causes.

The MODIS instruments are aboard both the Aqua and Terra satellites, launched on May 4, 2002 and December 18, 1999, respectively. The MODIS instruments collect data over the entire globe in 2 days. The AOD is retrieved using dark target methods in bands at 550, 670, 870, 1240, 1630, and 2130 nm, over ocean, and at 470, 550, and 670 nm over land. Other wavelengths are also used in the retrieval, for instance, short-wave infrared wavelengths for the land algorithm. Previous MODIS aerosol validation studies have compared the Aqua and Terra MODIS-retrieved AOD with the ground-based AERONET observations. AERONET is a global system of ground-based sun and sky scanning sun photometers that measure AOD in various channels, depending on individual instrument, but usually include measurements at 340, 380, 440, 500, 675, 870, and 1020 nm. Measurements are taken every 15 min during daylight hours. AERONET Level 2 quality assured AOD observations are accurate to within 0.01 for wavelengths of 440 nm and higher.

Previous studies concluded that MODIS AOD agreed with AERONET observations to within MODIS expected uncertainties, on a global basis. AERONET is only available for land locations, although some sites are in coastal regions. However, the correlation for the MODIS ocean algorithm was much better than the agreement for the MODIS land algorithm, in the Collection 4 data set. Revision and implementation of a new land algorithm and reprocessing of the data resulted in much improvement to the retrieved MODIS AOD over land. Even so, there remains a small overprediction of the AOD for low values and underprediction at high AOD values.

Data Description

Lary et al. (2009) used the global 10 km MODIS Collection 5 AOD product, over land and ocean, and all the available AERONET version 2.0 data. The AERONET program provides a long-term, continuous, and readily accessible public domain database of aerosol optical properties. The network imposes standardization of instruments, calibration, processing, and distribution. The location of individual sites is available from the AERONET web site http://aeronet.gsfc.nasa.gov/.

Lary et al. (2009) first identified all MODIS overpasses of the AERONET sites throughout the lifetime of the two MODIS missions. Then used the single green band MODIS AOD (550 nm) in the geographic grid point that contains the AERONET site. AERONET AOD measurements within 30 min of the MODIS observation are averaged. AERONET data are interpolated (in log–log space) to the green band where they are missing. They found a strong correlation between geographic location and bias. For example, there is a negative bias (MODIS under-estimation relative to AERONET) over vegetated Western Africa (from Liberia to Nigeria) and a positive bias over the Southwestern U.S. The spatial dependence of the differences between AERONET and MODIS is shown in Fig. 7.

Fig. 7 MODIS bias with respect to AERONET. Computed as a regression with intercept at the origin. Red indicates that MODIS is higher; blue indicates that AERONET is higher. The size of the circle is proportional to the slope of the regression for slope > 1 (where MODIS is higher) and to the inverse of the slope for slope < 1

Machine Learning AOD Bias Correction

Lary et al. (2009) applied two types of machine learning to the correction of the bias between MODIS and AERONET, i.e., neural networks and support vector machines (SVMs). For each of these machine-learning approaches, they used two training data sets, i.e., one for MODIS Aqua and one for MODIS Terra. These training data sets include all contemporaneous measurements of the MODIS instruments and AERONET made from launch to the present that were within 30 min of each other, within a great circle distance of 0.25°, and within a solar zenith angle of 0.1°. For MODIS Aqua, this gave a training record of 7543 points, and for Terra, 13,034 points.

The purpose of training a machine-learning algorithm is to construct a mapping between a set of input variables and an output variable (i.e., a multivariate nonlinear nonparametric fit). For each data set, the inputs were the surface type, the solar zenith angle, the solar azimuth angle, the sensor zenith angle, the sensor azimuth angle, the scattering angle, the reflectance, and the MODIS AOD. For each data set, the output was the AERONET AOD at 550 nm.

Figure 8c and d shows the result of performing a neural network bias correction. We see that the neural network is able to make a substantial improvement in the correlation coefficient with AERONET: an improvement from 0.86 to 0.96 for MODIS Aqua and an improvement from 0.84 to 0.92 for MODIS Terra.

Figure 8e and f shows the result of performing an SVR bias correction. The SVR makes an even greater improvement than the neural network correction, improving the correlation coefficient from 0.86 to 0.99 for MODIS Aqua and from 0.84 to 0.99 for MODIS Terra.

Examining the linear regression on the SVM fit, we see that the intercept (bias) is considerably reduced, from 0.03 to 0.0005 for Aqua and from 0.03 to 0.0001 for Terra. In addition, the slope of the SVM fit is almost 1 (0.99) for both Aqua and Terra.

Overall, the machine-learning results of Lary et al. (2009) (Fig. 8) show us that there is opportunity in the MODIS aerosol algorithm to improve the accuracy of the AOD retrieval, as compared with AERONET, and that this improvement is linked to surface type. We can use information from AERONET, from other satellite sensors such as MISR, and from detailed field experiments to continue to test and refine the assumptions in the MODIS algorithm. The results from the machine-learning analysis that point to surface type as the missing piece of information will allow us to focus the refinement procedure where it will help most.

Machine-learning algorithms were able to effectively adjust the AOD bias seen between the MODIS instruments and AERONET. SVMs performed the best, improving the correlation coefficient between the AERONET AOD and the MODIS AOD from 0.86 to 0.99 for MODIS Aqua and from 0.84 to 0.99 for MODIS Terra. Key in allowing the machine-learning algorithms to "correct" the MODIS bias was provision of the surface type and other ancillary variables that explain the variance between the MODIS and AERONET AOD.

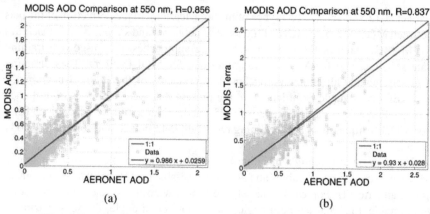

(a)   (b)

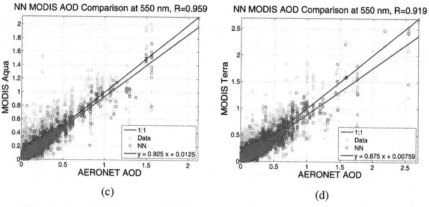

(c)   (d)

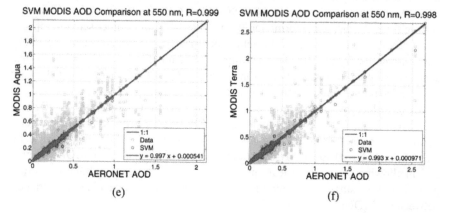

(e)   (f)

Fig. 8 (continued)

Machine Learning for New Product Creation

Let us know turn our attention to an example of creating new data products through the holistic use of satellite and in situ data. A new data product that is of societal significance.

Airborne Particulates

There is an increasing awareness of the health impacts of particulate matter and a growing need to quantify the spatial and temporal variations of the global abundance of ground level airborne particulate matter ($PM_{2.5}$). In March 2014, the World Health Organization (WHO) released a report that in 2012 alone, a staggering 7 million people died as a result of air pollution exposure (), one in eight of the total global deaths. A major component of this pollution is airborne particulate matter (e.g., $PM_{2.5}$ and PM_{10}).

The recent study by Lary et al. (2014) used machine learning to provide daily global estimates of airborne $PM_{2.5}$ from 1997 to 2014. This was achieved utilizing by using a massive amount of data (40 TB) from a suite of about 100 remote sensing and meteorological data products together with ground based observations of $PM_{2.5}$ from 8329 measurement sites in 55 countries taken between 1997 and 2014. This data was used to train a machine learning algorithm to estimate the daily distributions of $PM_{2.5}$ from 1997 to 2014. This allowed the creation of a new global $PM_{2.5}$ product at 10 km resolution from August 1997 to present (Lary et al., 2014). This new dataset is specifically designed to support health impact studies. Lary et al. (2014) showed some examples of this global $PM_{2.5}$ dataset and finish by examining a mental health Emergency Room admissions in Baltimore, MD. They demonstrate that the new $PM_{2.5}$ data product can reliably represent global observations of $PM_{2.5}$ for epidemiological studies. They showed that airborne particulates can have some surprising associations with health outcomes. As an example of this, Lary

Fig. 8 Scatter diagram comparisons of AOD from AERONET (*x*-axis) and MODIS (*y*-axis) as green circles overlaid with the ideal case of perfect agreement (blue line). The measurements shown in the comparison were made within half an hour of each other, with a great circle separation of less than $0.25°$ and with a solar zenith angle difference of less than $0.1°$. The left-hand column of plots is for MODIS Aqua, and the right-hand column of plots is for MODIS Terra. The first row shows the comparisons between AERONET and MODIS for the entire period of overlap between the MODIS and AERONET instruments from the launch of the MODIS instrument to the present. The second row shows the same comparison overlaid with the neural network correction as red circles. We note that the neural network bias correction makes a substantial improvement in the correlation coefficient with AERONET. An improvement from 0.86 to 0.96 for MODIS Aqua and an improvement from 0.84 to 0.92 for MODIS Terra. The third row shows the comparison overlaid with the SVR correction as red circles. We note that the SVR bias correction makes an even greater improvement in the correlation coefficient than the neural network correction. An improvement from 0.86 to 0.99 for MODIS Aqua and an improvement from 0.84 to 0.99 for MODIS Terra

et al. (2014) presented an analysis of Baltimore schizophrenia Emergency Room admissions in the context of the levels of ambient pollution. $PM_{2.5}$ had a statistically significant association with some aspects of mental health.

A useful validation of the new $PM_{2.5}$ data product is to survey the key features of the global $PM_{2.5}$ distribution and see if they capture what we expect to find and what has been reported in the literature. In Fig. 9a we see that the eastern half of the USA has a higher average abundance of $PM_{2.5}$ than the western half of the USA with the exception of California. This is consistent with the overlaid EPA observations shown as color filled circles. The color fill for the observations uses the same color scale as the machine learning estimate depicted using the background colors. There are persistently high levels of $PM_{2.5}$ in Mexico's dusty and desolate Baja California Sur. The particularly high values are in Mulegé Municipality close to Guerrero Negro (marked A in panel (a) of Fig. 9). Straddling the region close to the Mexico, Arizona, and California borders is the Sonoran Desert. This is a region characterized by a high average $PM_{2.5}$ abundance (marked B) and haboobs, massive dust storms. The Sonoran desert has an area of 311,000 square kilometers and is one of the hottest and dustiest parts of North America. This is clearly evident in the high 16-year average $PM_{2.5}$ abundance in this region. The persistently high $PM_{2.5}$ abundance associated with Los Angeles is visible (marked C). The regions of high population density usually coincide with the region of high particulate abundance. California's heavily agricultural Central Valley has a high $PM_{2.5}$ loading (marked D), note the good agreement of our estimates with the 16 year average observations. The EPA has designated Central Valley as a non-attainment area for the 24-h $PM_{2.5}$ National Ambient Air Quality Standards (NAAQS). The high $PM_{2.5}$ abundance associated with the Great Salt Lake Desert in northern Utah close to the Nevada border is clearly visible (marked E). There is a nearby measurement supersite at Salt Lake City recording a particulate abundances consistent with our estimates. Mexico City is known for its high levels of particulates and is clearly visible (marked F) as a localized hot spot. Close to the Mexico/Texas border we see the elevated $PM_{2.5}$ abundance associated with the Chihuahuan Desert and the Big Bend Desert (marked G). Dust storms in this area often impact El Paso in Texas and Ciudad Juarez in Mexico. The Ohio River Valley (marked H) encompasses several states and is home to numerous coal-fired power plants, chemical plants, and industrial facilities, leading to high levels of ambient particulates. The Ohio River Valley has a higher average abundance of $PM_{2.5}$ than the rest of the East Coast. Our analysis agrees closely with the in-situ observations for the Athens super-site. The Piura desert in Northern Peru (marked I) on the coast and western slopes of the Andes is a region of high particulate abundances. The region in South America from the high Andean semi-arid Altiplano basin in the north, coming down through the Salar de Uyuni Desert (the world's largest salt flats), passing by Santiago in Chile and San Miguel de Tucumn, San Juan and Mendoza in Argentina, and down to the Neuquén Basin in the south is characterized by a high abundance of particulates from a combination of dust, salt, and pollution (marked J). The southern Amazon in Bolivia and the surrounding region has a lot of burning leading to persistently high particulate abundances (labeled K).

PM$_{2.5}$ Multi-Year Average 1997-2013 (5874 days)

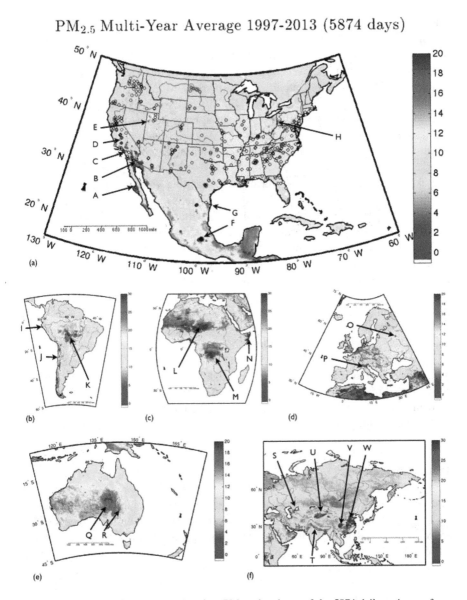

Fig. 9 The average of the estimated surface PM$_{2.5}$ abundance of the 5874 daily estimates from August 1, 1997, to August 31, 2013 in μg/m^3 for (**a**) the USA, (**b**) South America, (**c**) Africa, (**d**) Europe, (**e**) Australia, and (**f**) Asia

The Bodélé depression is Chad's lowest point on the Sahara's southern edge that supplies the Amazon forest with the majority of its mineral dust. The high abundance of PM$_{2.5}$ over the Bodélé is clearly visible (marked L). Typically there are dust storms originating from the Bodélé depression on around 100 days a year.

The low flat desert in the North African Western Sahara is some of the most inhospitable and arid land on earth and a substantial dust source, clearly visible in the high abundance of $PM_{2.5}$. Burning in the Democratic Republic of the Congo (marked M) leads to high levels of particulates. Much of coastal Somalia is desert characterized by high levels of particulates (marked N).

The Italian Po valley (marked P in Fig. 9) has some of the highest average abundance of particulates in Europe. Industrial emissions coupled with persistent fog leading to heavy smog. High levels of $PM_{2.5}$ are found in the Netherlands and North-west Germany. An example of a local pollution hotspot in Europe is Moscow (marked O).

Lake Eyre is Australia's largest lake and lowest point (marked Q). When the lake has dried out a salt crust remains. When Lake Eyre is dry it is typically Australia's largest dust source, Lake Eyre usually only fills with water after the heavy rains that typically occur once every 3 years, during these periods the $PM_{2.5}$ abundance in the vicinity of Lake Eyre is lower than usual. Just east of the Lake Eyre Basin is the Strzelecki Desert another major Australian dust source (marked R). The arid region just south of the Hamersley Range in Western Australia, the Gibson Desert, Great Victoria Desert and MacDonnell Ranges are also dusty environments with elevated average abundances of $PM_{2.5}$.

Asia has some of the highest particulate abundances anywhere on earth. The Aral Sea (marked S) lying across the border of Kazakhstan and Uzbekistan is heavily polluted with major public health problems. The Ganges Valley is home to 100 million people and is highly polluted (marked T). The cold Taklimakan Desert of northwest China is a major source of $PM_{2.5}$ (marked U). Particularly high levels of particulates are found in the Sichuan Basin (marked V) and in western China in the region from Beijing in the North down to Guangxi in the south (marked W).

Tracer Correlations

The spatial distributions of atmospheric trace constituents are in general dependent on both chemistry and transport. Compact correlations between long-lived species are well observed features in the middle atmosphere. The correlations exist for all long-lived tracers—not just those which are chemically related—due to their transport by the general circulation of the atmosphere. The tight relationships between different constituents have led to many analyses where measurements of one tracer are used to infer the abundance of another tracer. These correlations can also be used as a diagnostic of mixing and to distinguish between air-parcels of different origins.

Of special interest are the so-called "long-lived tracers": constituents such as nitrous oxide (N_2O), methane (CH_4), and the chlorofluorocarbons (CFCs) that have long lifetimes (many years) in the troposphere and lower stratosphere, but are destroyed rapidly in the middle and upper stratosphere.

The correlations are spatially and temporally dependent. For example, there is a compact-relation regime in the lower part of the stratosphere and an altitude-dependent regime above this. In the compact-relation region, the abundance of one tracer is uniquely determined by the value of the other tracer, without regard to other variables such as latitude or altitude. In the altitude-dependent regime, the correlation generally shows significant variation with altitude.

The description of such spatially and temporally dependent correlations is usually achieved by a family of correlations. However, a single neural network is a natural and effective alternative.

Reconstructing CH_4–N_2O Correlations

The motivation for this study was preparation for a long-term chemical assimilation of Upper Atmosphere Research Satellite (UARS) data starting in 1991. For this period we have continuous version 19 data from the Halogen Occultation Experiment (HALOE) but not observations of N_2O as both ISAMS and CLAES failed. In addition we would like to constrain the total amount of reactive nitrogen, chlorine, and bromine in a self-consistent way (i.e., the correlations between the long-lived tracers are preserved). Tracer correlations provide a means to do this by using HALOE CH_4 observations.

Figure 10a shows the CH_4–N_2O correlation from the Cambridge 2D model overlaid with a neural network fit to the correlation (Lary et al., 2003). The neural network used was a feed-forward multilayer perceptron. There were four inputs, one output, and one hidden layer with eight nodes. A non-linear activation function was used. The training dataset contained 1292 patterns, sampling the input space completely as shown in Fig. 10. The network was constrained for 106 epochs (iterations).

The correlation coefficient between the actual solution and the neural network solution was 0.9995. Figure 10 panel (b) shows how the median fractional error of the neural network decreases with epoch (iteration). Both CH_4 and pressure are strongly correlated with N_2O as can be seen in panels (c) and (d). Latitude and time are only weakly correlated with N_2O as can be seen in panels (e) and (f). Even though the correlation with time of year and latitude is relatively weak it still does play a role in capturing some of the details of the CH_4–N_2O correlation in Panel (a).

A polynomial or other fit will typically do a good job of describing the CH_4–N_2O correlation for high values of CH_4 and N_2O. However, for low values of CH_4 and N_2O there is quite a spread in the relationship which a single curve can not describe. This is the altitude dependent regime where the correlation shows significant variation with altitude.

Figure 10c shows a more conventional fit using a Chebyshev polynomial of order 20. This fit was chosen as giving the best agreement to the CH_4–N_2O correlation after performing fits using 3667 different equations. Even though this is a good fit the spread of values cannot be described by a single curve. However, a neural network trained with the latitude, pressure, time of year, and CH_4 volume mixing ratio (v.m.r.) (four inputs) is able to well reproduce the N_2O v.m.r. (one output), including the spread for low values of CH_4 and N_2O.

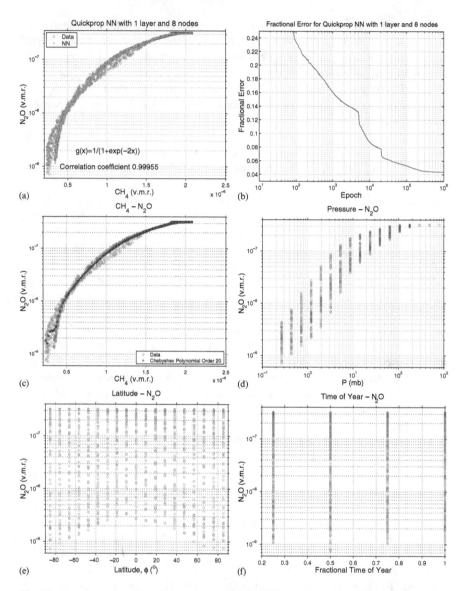

Fig. 10 The neural network used to produce the CH_4-N_2O correlation in Panel (**a**) used a neural network with one hidden layer with eight nodes. The correlation coefficient between the actual solution and the neural network solution was 0.9995. Panel (**b**) shows how the median fractional error of the neural network decreases with epoch (iteration). Both CH_4 and pressure are strongly correlated with N_2O as can be seen in panels (**c**) and (**d**). Latitude and time are only weakly correlated with N_2O as can be seen in panels (**e**) and (**f**). Even though the correlation with time of year and latitude is relatively weak it still does play a role in capturing some of the details of the CH_4-N_2O correlation in Panel (**a**)

Variable scaling often allows neural networks to achieve better results. In this case all variables were scaled to vary between zero and one. If the initial range of values was more than an order of magnitude, then log scaling was also applied. In the case of time of year the sine of the fractional time of year was used to avoid a step discontinuity at the start of the year.

Neural networks are clearly ideally suited to describe the spatial and temporal dependence of tracer-tracer correlations (Lary et al., 2003). Even in regions when the correlations are less compact. Useful insight can be gained into the relative roles of the input variables from visualizing the network weight assignment.

Pollen Estimation

Pollen is known to be a trigger for allergic diseases, e.g. asthma, hay fever, and allergic rhinitis (Oswalt and Marshall, 2008; Howard and Levetin, 2014). It is interesting that a variety of non-respiratory issues such as strokes (Low et al., 2006), and surprisingly, even suicide and attempted suicide (Matheson et al., 2008) have an association with the daily concentration of atmospheric particulates. However, so far, there is no defined threshold amount of pollen known to trigger allergy for sensitive individuals (Voukantsis et al., 2010). One of the factors for the lack of knowledge of the threshold amount of pollen is the absence of an accurate estimation on a fine spatial scale of the hourly, bi-hourly, or daily amount of pollen. Individual physiological differences such as gender and age among sensitive people also adversely affect in knowing the threshold amount of pollen in the surrounding (Britton et al., 1994; Ernst et al., 2002).

Of all plants, weeds, and particularly those of the *Ambrosia* species, e.g. *Ambrosia artemisiifolia* (common ragweed), *Ambrosia trifida* (giant ragweed) are major producers of large amounts of pollen. For example, a common ragweed can produce up to about 2.5 billion pollen grains per plant per day (Laaidi et al., 2003). *Ambrosia artemisiifolia* and *Ambrosia trifida* combined can produce more allergens than all other plants combined (Lewis et al., 1983). Grasses (e.g., *rye grass*) are also known to trigger an allergic response. Following *Ambrosia artemisiifolia*, grass pollen are known for their high allergic potency than most weeds (Esch et al., 2001; Lewis et al., 1983). Tree pollen can cause an allergic response, but one that is typically less than that of weeds and grasses, although in some regions tree pollen can trigger a significant allergic response. For instance, the airborne concentration of Mountain cedar pollen grains can reach tens of thousand of pollen grains per cubic meter and trigger a significant allergic response in central Texas during winter, known as cedar fever (Andrews et al., 2013; Ramirez, 1984).

Both global climate change and air pollution affect the abundance of airborne pollen, and consequently, its allergic impact (Kinney, 2008; Wayne et al., 2002; Voukantsis et al., 2010). For example, the abundance of pollutants such as CO_2, Wayne et al. (2002) and NO_2 (Zhao et al., 2016) can affect the extent of growing season of major pollen producing plants, and thereby also affect the airborne pollen concentration as well as altering the onset and end dates of seasonal allergies.

Overall, more people are exposed to pollen and sensitive individuals become exposed to large amount of pollen for longer period of time over larger areas.

Globally millions of people are affected by seasonal allergies, and the number of people affected is increasing each year. In North American alone, as of 2008, about 50 million adult Americans and 9% of children aged below 18 have experienced pollen caused allergies (Howard and Levetin, 2014). Similarly, in Europe about 15 million people are affected by hay fever, asthma, and rhinitis (D'amato and Spieksma, 1991). Hence, pollen allergies are becoming an increasingly significant environmental health issue. Hence, just as accurate daily weather forecasts are of significant use, accurate daily pollen forecasts are likely to become increasingly important.

Remote Sensing has been employed to study atmospheric pollen concentrations. For example, the polarization of LIDARs has been used to observe the airborne tree pollen abundance at Fairbanks Alaska (Sassen, 2008). In this case, the pollen produces a depolarization of the LIDAR backscattering signals from the lower atmosphere. The light scattering properties of pollen are also manifested in the shape of the solar corona they create. The shape of the solar corona associated with pollen depends on the shape of the pollen grains and their atmospheric concentration (Tränkle and Mielke, 1994). However, this approach can be complicated as atmospheric light scattering is also caused by other airborne particulates.

Common pollen estimation techniques, particularly those made in Europe, stress the importance of meteorologic variables (Kasprzyk, 2008). Usually forecasting the amount of airborne pollen is based on the interaction of atmospheric weather and pollen (Arizmendi et al., 1993). Meteorologic variables such as the daily mean, maximum, change in temperature and dew point variables show positive correlation with the pollen concentration (Kasprzyk, 2008). Kasprzyk (2008) found that atmospheric humidity shows negative correlation to the pollen concentration. Other studies show that temperature, precipitation, and wind speed are significant meteorologic parameters in estimating pollen concentration (Stark et al., 1997).

Most of these meteorologic variable based forecasting methods employed statistical methods such as linear regression, the polynomial method and time serious analysis (Sánchez-Mesa et al., 2002). Only few studies used advanced machine learning methods such as neural network (Sánchez-Mesa et al., 2002; Rodríguez-Rajo et al., 2010; Puc, 2012; Voukantsis et al., 2010) and random forest (Nowosad, 2016) for pollen forecasting and support vector machines are applied for related environmental studies (Voukantsis et al., 2010; Osowski and Garanty, 2007).

Predicting Pollen Abundance

Over the past decade neural networks have been applied to study pollen of different species over the European region. For example, Csépe et al. (2014) used different Computational Intelligence (CI) methods to predict the *Ambrosia* pollen at two different places in Hungary and France. Castellano-Méndez et al. (2005) and Puc (2012) have employed the neural network to predict *Betula* pollen over Spain and

Table 1 Name and type of predictors (input variables) used for our machine learning training. Parameters consist of environmental and NEXRAD radar measurements

Parameter	Unit	Type
Vegetation greenness fraction	Fraction	Env.
Leaf area index	m^2	Env.
Roughness length, sensible heat	m	Env.
Displacement height	m	Env.
Energy stored in land	Jm^{-2}	Env.
Mean reflectivity	dB	NEXRAD
Mean Doppler velocity	ms^{-1}	NEXRAD
Mean spectral width	ms^{-1}	NEXRAD
Reflectivity [10–10 dB]	dB	NEXRAD
Velocity [10–10 dB]	ms^{-1}	NEXRAD
Spectral width [10–10] dB	ms^{-1}	NEXRAD
Reflectivity [20–20 dB]	dB	NEXRAD
Velocity [20–20 dB]	ms^{-1}	NEXRAD
Spectral width [20–20 dB]	ms^{-1}	NEXRAD
Reflectivity [40–40 dB]	dB	NEXRAD
Velocity [40–40 dB]	ms^{-1}	NEXRAD
Spectral width [40–40 dB]	ms^{-1}	NEXRAD
Wind direction at altitude 50 m	Degree	NEXRAD
Wind speed at altitude 50 m	ms^{-1}	NEXRAD

Poland, respectively. Recently, Nowosad (2016) used the random forest method to forecast different tree pollen species.

In this study we used random forests, neural networks, and support vector machines to estimate daily *Ambrosia* pollen concentration at Tulsa, Oklahoma (location: 36.1511°N, 95.9446°W). We used a combination of environmental parameters and NEXRAD radar measurements. The combined parameters are listed in Table 1. The daily pollen concentration used in the training of our machine learning algorithms was obtained using a Burkhard spore trap at the University of Tulsa, Oklahoma.

After pollen is produced in the plant anthers its emission dispersion and deposition is influenced by meteorological variables such as the temperature, wind speed, and direction and pressure (Kasprzyk, 2008; Csépe et al., 2014; Howard and Levetin, 2014). Other meteorologic parameters such as dew point, humidity, rainfall sunshine duration are also known to affect pollen emission and distribution (Kasprzyk, 2008).

We used a set of environmental and NEXRAD radar parameters (Table 1) in our machine learning training. Environmental parameters such as vegetation greenness fraction, roughness length (sensible heat), energy stored in all land reservoirs, and displacement height and leaf area index are selected. The other set of data we used are the NEXRAD measurements which consist of the reflectivity, Doppler velocity, and spectral width which represent, respectively, the amount of back scattered signals from a scattering volume, the velocity of the scatterer along the radar line of sight and the width of the power spectrum. All NEXRAD measurements are taken at the lowest elevation. Additionally the NEXRAD provides measurements of the vertical profile of the direction and speed of the wind from about near the surface of the Earth. The dual polarization measurements: differential reflectivity, differential phase, and correlation coefficient use the horizontal and vertical polarization signals and are particularly suited for particle identification. In this study we do not use the dual polarization (polarimetric) NEXTRAD measurements as we have only few days of the measurements in contrary to the ideal high dimensional data requirement for machine learning.

The three machine learning methods were trained on the entire data set to assess their performance in predicting the *Ambrosia* pollen. The scatter diagrams are shown in Fig. 11. We also used a Newton-Raphson based recursive Random Forest technique that has been developed in order to improve the accuracy. The method includes error estimation and correction. In order to evaluate the performance of the machine learning methods independently, 10% of the data are randomly selected and withdrawn for validation from the training and the remaining 90% of the data is then used for training the model. After developing the model, its performance is tested using the independent 10% of the validation dataset that was not used in training the machine learning regression. These results are shown in Fig. 11. Panels (a)–(c) in Fig. 11 show scatter plots of prediction made by the support vector machine, neural network, and random forest machine learning methods, respectively, using

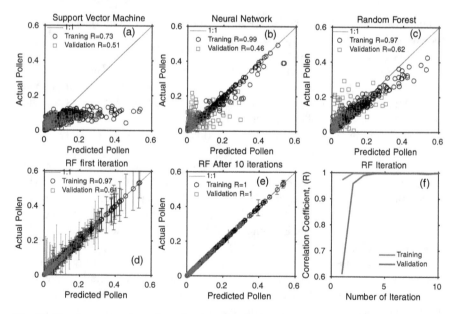

Fig. 11 Showing scatter plots of actual and predicted pollen for the support vector machine (panel **a**), neural network (panel **b**), random forest (panel **c**). Panel (**d**)–(**f**) shows results of an iteration method applied to the random forest. Panel (**d**) and (**e**) shows results of the first iteration 10th iterations for the training (black circles) and validation data (red squares). Panel (**f**) depicts plots of the correlation coefficient for the training and validation data versus iteration number

the training data (black circles) and the validation data (red squares). Results of the iterative method applied to the random forest method are given by panels (d)–(f). The random forest machine learning is trained using 200 decision trees.

From the top three panels of Fig. 11, we observe that the neural network and random forest methods produced better predictions than the Support Vector machine. The random forest method produced the best independent validation results (correlation coefficient, 0.62) of all the three methods. The high correlation value of neural network found using the training data (correlation coefficient 0.99) is not reproduced in the independent validation test which had a correlation coefficient of only 0.46. Error bar plots for the training and the validation data for the first iteration of the random forest are given by panel (d) in Fig. 11. We see that predictions using both the training and validation data exhibit large errors and a low correlation coefficient. Interestingly, after a few iterations the random forest produced results with significantly reduced errors and correlation values close to 1 (panel (e) in Fig. 11). Panel (f) in Fig. 11 shows the correlation coefficient values between the normalized estimated and actual pollen for the training (blue curve) and validation data (red curve) sets for 10 iterations. We observe that the iterative of the random forest method has reduced the error significantly and the correlation coefficient values converge to one for both training and validation data sets.

The upper panel of Fig. 12 shows a comparison of the actual and predicted pollen using the recursive random forest. Another important application of machine learning methods is the selection of the best features (variables) that contribute most to the prediction and ranking them in order of the importance. In this way we can determine the most important predictor variables and estimate the output leaving features that contribute less. The random forest provides such a ranking based on criteria attributed to the splitting variable in the data sampling to form decision tree (Genuer et al., 2010; Kotsiantis et al., 2007; Friedman et al., 2001).

The lower panel of Fig. 12 shows the ranking of the relative importance of the variables provided by the random forest with 200 trees. The most important factors in estimating the pollen were the leaf area index, vegetation greenness function, and displacement height.

Using Machine Learning for Ocean Data Products

Let us take a look at using machine learning to estimate chromophoric dissolved organic material (CDOM) absorption. A quality control database has recently been assembled (Aurin and Mannino, 2012) of optical and biogeochemical parameters suitable for supporting the development of ocean color algorithms to retrieve chromophoric dissolved organic material (CDOM) absorption (a_g), CDOM spectral slope (S_g), and Dissolved Organic Carbon (DOC) from satellite-derived, multi-spectral remote sensing reflection (Aurin and Mannino, 2012). This database of sea-surface CDOM is considerably larger (18,035 stations) than NOMADv2 (a subset of SeaBASS that is often used in algorithm development; 1182 stations)

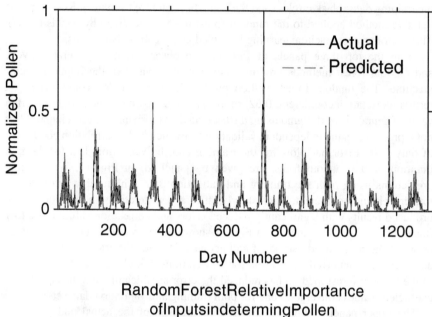

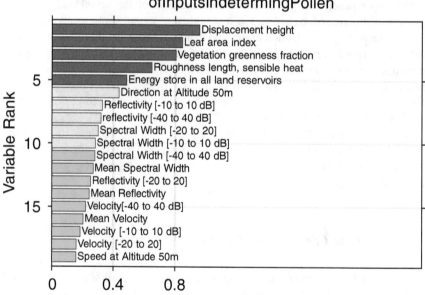

Fig. 12 The upper panel shows the comparison of actual and predicted pollen time series for Tulsa, OK. The lower panel shows the ranking of the relative importance of the variables provided by the random forest with 200 trees

with valid satellite matches at 8654 stations. In the process, estimation of S_g at multiple wavebands between 245 and 715 nm was found to be insensitive to the spectral resolution of a_g, but did depend on reference waveband selection. Global distributions of a_g and S_g are analyzed for patterns related to source, composition, and photo-degradation of CDOM.

CDOM absorbs light most strongly in the ultraviolet and blue region of the spectrum. Non-turbid waters with no CDOM appear blue and as the CDOM concentration increases the color of the water will change from green, through yellow-green to brown. The increasing levels of CDOM diminish light penetration and hence affect the biological activity of aquatic systems by limiting photosynthesis and inhibiting the growth of phytoplankton. CDOM also helps protect organisms from DNA damage by absorbing the harmful UVA and UVB radiation.

Figure 14 shows the global CDOM estimate obtained when the Random Forest multivariate non-parametric fit is used together with a global 9 km SeaWIFS Mapped Seasonal multi-wavelength radiances. The CDOM distribution shown in Fig. 14 is realistic, with low CDOM in regions such as the South Atlantic and South Pacific Gyres, and high CDOM in freshwater and in coastal areas with river outflows.

CDOM abundance is variable, typically with the lower abundances in the open ocean and higher abundances in fresh waters and estuaries. The database (Aurin and Mannino, 2012) was used together with machine learning to develop a set of algorithms to estimate a_g and S_g based on the multi-wavelength remote sensing reflectance observed by SeaWIFS, MODIS Aqua, and MODIS Terra.

Inputs The training dataset used provides training examples to learn from of the multi-wavelength radiances observed by the satellite and of the corresponding ocean parameters such as CDOM. For SeaWIFS the wavelengths used are 412, 443, 490, 510, 555, and 670 nm. For MODIS Terra and Aqua the wavelengths used are 412, 443, 488, 531, 547, and 667 nm. In this study we only consider training examples that were made by the satellites at zenith angles of less than 60°, within 3 h of the in-situ observations, and with a coefficient of variability of less than 0.15.

Fit Quality Figure 13a shows in blue a scatter diagram of the training SeaWIFS CDOM values on the x-axis versus the random forest estimates on the y-axis. The red points show the independent validation dataset. There is a good clustering around the 1:1 line shown in green. The training provides a good fit with an R^2 value of 0.91. The independent verification with the validation dataset has an R^2 value of 0.72 (Table 2). Figure 13b shows the corresponding Taylor diagram, point A is the training observations, point B is the Random Forest fit.

A very useful feature of Random Forests is that they can provide a ranked list of the relative importance of the variables used in producing the regression. This is done by looking at the increase in the mean squared error averaged over all the trees in the ensemble and divided by the standard deviation taken over the trees, for each variable. The larger this value, the more important the variable is in producing a good quality fit. Figure 13c shows the relative importance of the Random Forest inputs. It indicates that in estimating CDOM at 412 nm using SeaWIFS the order of

Fig. 13 Panel (**a**) shows in
blue a scatter diagram of the
training SeaWIFS CDOM
values on the *x*-axis versus
the estimated values on the
y-axis. The red points show
the independent validation
dataset. Panel (**b**) shows the
corresponding Taylor
diagram, point A is the
training observations, point B
is the Random Forest fit.
Panel (**c**) shows the relative
importance of the Random
Forest inputs

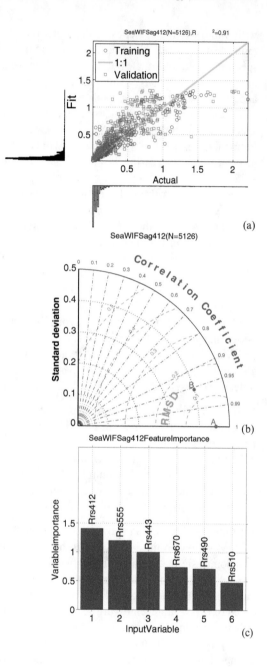

Table 2 Statistics for the SeaWIFS CDOM training

	σ	ϵ	R^2	n
Training observations	0.15			5126
Training fit	0.13	0.046	0.91	5126
Validation observations	0.33			569
Validation fit	0.35	0.19	0.72	569

Fig. 14 The global a_{CDOM} (412 nm) estimate obtained when the Random Forest multivariate non-parametric fit is used together with a global 9 km SeaWIFS Mapped Seasonal multi-wavelength radiances

Fig. 15 Dust sources are typically localized point sources

importance of the radiances are 412 nm > 555 nm > 443 nm > 670 nm > 490 nm > 510 nm.

Example Global Distribution Figure 14 shows the global CDOM estimate obtained when the Random Forest multivariate non-parametric fit is used together with a global 9 km SeaWIFS Mapped Seasonal multi-wavelength radiances. The CDOM distribution shown in Fig. 14 is realistic, with low CDOM in regions such as the South Atlantic and South Pacific Gyres, and high CDOM in freshwater and in coastal areas with river outflows.

Dust Source Identification Using Unsupervised Classification

Unsupervised classification can be very useful when we would like to objectively split up our data into different regimes. A good example of this is a study to characterize dust sources (e.g., Fig. 15) (Lary et al., 2016).

Dust sources of many kinds are found globally. One of the most salient features of dust sources is that they are often very localized. For example, in Figs. 15 and 17 we can clearly see that the source of the dust plumes is best described as an ensemble of many point sources, not broad dust emitting regions. Realistically capturing this very localized nature of dust sources has so far largely eluded automated diagnosis, and consequently, description in global models. Invariably current models describe dust sources as rather large scale features, even when vegetation indices and similar approaches are used. This is in marked contrast to what we consistently see in the satellite imagery across the planet (e.g., Figs. 15 and 17).

Identifying dust sources is a critical yet challenging task for the accurate simulation of atmospheric particulate distributions relevant to air quality and climate change.

We take a new and radically different approach to any previous studies that have sought to identify global dust sources on a routine basis. We demonstrate that this new approach employing machine learning is very effective. The approach uses multi-wavelength spectral reflectivity signatures to characterize land surfaces, naturally paving the way for a new class of algorithm ideally suited to fully exploit the next generation of hyper-spectral instrument. The production of thematic maps, such as those depicting land cover, using an image classification is one of the most common applications of remote sensing. New in our approach is that we can both operate at very high spatial resolution and distinguish between types of dust sources. For example, we can easily distinguish between the edge of salt flats (Fig. 17), dried up wadis or lakes, and agricultural sources to name just three of many examples. The only limiting factor for the resolution is the resolution of the satellite imagery.

We employ machine learning to objectively provide an unsupervised multi-variate and non-linear classification into a very large number of surface types (in our demonstration study presented below 1000 classes are used) using multi-spectral satellite data. In other words, we do not impose any a priori assumptions, but rather, we let the data speak for itself as to how we should classify surface types. Self-organizing maps (SOMs) are a data visualization and unsupervised classification technique invented by Professor Teuvo Kohonen that reduce the dimensions of data through the use of self-organizing neural networks.

SOMs help us address the issue that humans simply cannot visualize high dimensional data unaided. The way SOMs go about reducing dimensionality is by producing a feature map, usually with two dimensions, that objectively plots the similarities of the data by grouping similar data items together. SOMs learn to classify input vectors according to how they are grouped in the input space. The SOM learns to recognize neighboring sections of the input space. Thus, SOMs learn both the distribution and topology of the input vectors they are trained on.

This approach allows SOMs to display similarities and reduce the dimensionality. A SOM does not assume a priori a functional form for the analyzed data. A noteworthy enhancement of an SOM over principal component analysis is an SOM's ability to represent non-linear functions or mappings.

The premise being that there are very many types of dust sources, from the diatom rich sediments of the Bodélé depression in Chad, to those at the edge of salt flats in Bolivia and Chile (Fig. 17), to those in the coastal Green Mountains of Libya. Each of these dust sources has distinct physical characteristics, and therefore a distinct reflectance signature. If we are able to identify these signatures, then we can map the temporal and spatial evolution of each of these distinct dust sources. Once we have the surface type classification we then seek to identify which small subset of surface classes correspond to various kinds of dust sources. Once we have identified the signature of a wide variety of dust sources we can precisely pick out these locations globally and how their distribution changes with time. This is particularly useful as dust sources are very localized, and some dust sources have a significant seasonal time evolution. Having a methodology to identify the signature of these small-scale regions is invaluable.

The machine learning approach to dust source identification was first conceived in 2010 to face a very practical challenge that the Navy has in producing real time visibility forecasts. If the standard type of dust sources are used it was found that very poor regional visibility forecasts result. However, the quality of the Navy visibility forecasts drastically improved with an analyst (Annette Walker) manually identifying individual dust sources at the heads of plumes by examining sequences of satellite images such as those shown in Fig. 17 and also the EUMETSAT RGB Composites Dust images available online (http://oiswww.eumetsat.org/IPPS/html/ MSG/RGB/DUST/). This methodology is very labor intensive and does not lend itself to easy automation. The first prototype dust sources using the machine learning approach described here were devised specifically to automate the dust source identification and also allow for the accurate diagnosis of the time evolution in the spatial extent of the dust sources. Beyond the applications of accurate dust sources for visibility and air quality forecasts, the radiative forcing (RF) due to dust is a key concept in climate change calculations considered by the IPCC for the quantitative comparison of the strength of different human and natural agents causing climate change. Radiative forcing can be categorized into direct and indirect effects. A significant part of the direct effect is the mechanism by which aerosols scatter and absorb shortwave and longwave radiation, thereby altering the radiative balance of the Earth—atmosphere system. Mineral dust is a major component of global aerosols that exert a significant direct radiative forcing. Mineral dust aerosols are produced both naturally ($\approx 70\%$) and anthropogenically ($\approx 30\%$).

Our ultimate goal is to identify all the surface locations on the planet that are dust sources. To do this we use a SOM to classify all the land surface locations into a very large set of n categories. In the examples shown here, $n = 1000$. A small subset of these 1000 categories will be regions that are dust sources. Naturally, there are a variety of distinct types of dust sources (e.g., dry river beds, agricultural sources, edge of salt flats, etc.) that we would like to delineate.

Self Organizing Map Classification

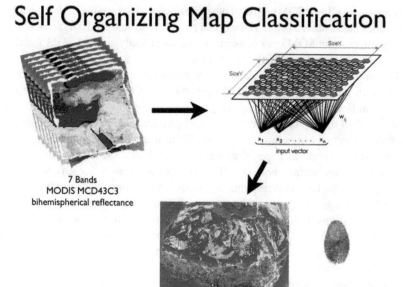

Fig. 16 Schematic of how self-organizing maps have been used in this study to classify land surface pixels into 1000 classes. Then a small subset of these classes are identified as dust sources

To achieve a comprehensive classification we want to consider the conditions present throughout the year, so in the demonstration we took an entire year of the 0.05° resolution MCD43C3 data product (Fig. 16). For this entire year of data, we then calculate the mean, μ for each grid point. This is a massive dataset, and the computational time and memory required to perform the SOM classification increases with the number of data records. For the examples shown here we therefore first restricted our attention to those broad MODIS surface types that may include dust sources, namely: barren or sparsely vegetated surfaces, croplands, grasslands, and open and closed shrublands. These are MODIS surface types 16, 12, 10, 7, and 6, respectively. For each of these surface types we then constructed an input vector that contains seven values, namely for each of the seven bands provided in the MCD43C3 MODIS product the mean, μ, of the directional and bihemispherical reflectance. When training the SOM we use the Euclidean distance to compare the input vectors (each containing seven values).

In order to provide a fine gradation of classification we use the SOM to group together the surface locations into 1000 classes, only a small subset of which correspond to regions that are dust sources. Once the classes that correspond to dust sources have been successfully identified, we have an automated method with which we can identify dust sources that can be routinely executed to provide a regular dust source data product that captures the spatial and temporal evolution of dust sources globally. We utilized the extensive hand classification of very localized dust sources produced by the Navy for the Middle East and South West Asia to guide our initial

July 18, 2010 MODIS Aqua True Color

South America: Bolivia and Chile

Fig. 17 Example of our machine learning approach correctly identifying very localized point sources around the edge of salt flats in Bolivia and Chile. Notice the narrow dust plumes originating from precisely the identified source regions that have been highlighted in blue and cyan

determination of which of the 1000 classes are dust sources. It is worth noting that the SOM classes are unique and distinct, this will be seen below with the example of the Bodélé Depression. Classes near each other are similar, but distinct.

Bolivia and Chile Salt Flats Dust Event

Figure 17 shows the dust event of July 18, 2010 in the Bolivian Altiplano. This event can be seen clearly in the MODIS Aqua True Color image where dust plumes emanate from fluvio-lacustine deposits and fluviodeltaic sediments around

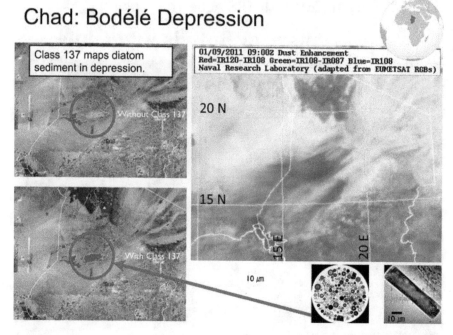

Chad: Bodélé Depression

Class 137 maps diatom sediment in depression.

01/09/2011 09:00Z Dust Enhancement
Red=IR120-IR108 Green=IR108-IR087 Blue=IR108
Naval Research Laboratory (adapted from EUMETSAT RGBs)

Without Class 137

With Class 137

20 N

15 N

15 E

20 E

10 μm

10 μm

Fig. 18 The Bodélé depression dust event of January 9, 2011 (7Z to 18Z). The right panel shows the NRL processed EUMETSAT MSG/RGB satellite product for January 9, 2011 09Z. The two left panels show the dust sources identified by our approach with (lower) and without (upper) SOM Class 137. Lower right insets show microscopic images of Bodélé diatoms from soils samples taken from the depression

the Salars de Coipasa and Uyuni, Lake Poopo, and other smaller salt flats and lakes. Overlaid are the SOM classes that coincide with active dust sources on the Altiplano. Notice that the salt flats themselves are not dust sources, rather we see the plumes forming around the edges of the flats and lakes. SOMs are very successful in identifying the unique spectral signatures of dust sources. A set of papers which is in preparation will be describing an exhaustive atlas of the global dust sources.

Bodélé Depression Dust Event

Figure 18 shows the Bodélé depression dust event of January 9, 2011 at 09Z that started at 7Z and ceased at 18Z. The Bodélé depression is Chad's lowest point on the Sahara's southern edge. Typically there are dust storms originating from the Bodélé depression on around 100 days a year that supplies the Amazon forest with the majority of its mineral dust (Washington and Todd, 2005; Koren et al., 2006; Washington et al., 2006; Todd et al., 2005; Bouet et al., 2005). The right panel shows the NRL processed EUMETSAT MSG/RGB satellite product. The two left panels show the dust sources identified by our approach with (lower) and without (upper) SOM class 137. The SOM had automatically determined that the sediment in the Bodélé depression was distinct from the surrounding dust sources and put it in a class all of its own, class 137. Indeed it is different, the Bodélé depression was once filled with a freshwater lake that has long since dried up (Washington et al., 2005). This has left behind diatoms that now make up the surface of the depression. The two key points being, first that the dust source of the Bodélé is distinct from the surrounding dust sources, and second, that it consists of diatoms. This is interesting as if we could devise a way of distinguishing dust sources with containing certain biological materials it would have significant applications for public health issues.

Characterizing Pelagic Habitats Within Coastal Waters

Commercial and recreational fisheries within the Gulf of Mexico contribute significantly to the region's ocean economy making effective management a priority. The goal of fisheries management is to optimize the benefits of living marine resources by addressing threats to a resources' sustainability through conservation, development, and full utilization of the fishery resources to provide food, employment, income, and recreation. Therefore, it is desirable to minimize management actions that may result in negative impacts on fishermen and the coastal community. However, depending on the source of the threat, some management actions such as implementing fishing gear restrictions, time and area closures, and harvest limits, may have direct adverse impacts on fishermen. Often issues impacting fisheries populations arise from degradation or loss of habitat, requiring a different management approach. Coastal and marine habitats can be significantly and rapidly impacted by a number of anthropogenic actions and natural events such as coastal storms, development and hydrological alterations. With approximately 98% of Gulf of Mexico fisheries dependent on estuarine and near-shore habitats at some point in their life cycle (Lellis-Dibble et al., 2008), it is critical that resource managers have the ability to quickly and frequently monitor and assess habitat loss and degradation. However, inconsistencies in the approach that various agencies use for naming habitats make it difficult to develop a region-wide habitat map without standardizing the information.

Unlike most habitat classification systems currently in use, the Coastal and Marine Ecological Classification Standard (CMECS) (Committee et al., 2012) has a water column component which identifies key classifiers required to characterize pelagic habitat types. The vastness and dynamic nature of the ocean's water column limit the feasibility of the frequent in situ sampling that would be necessary to monitor these classifiers and routinely produce region-wide map products. Our ultimate goal is to provide an example of how the Machine Learning classification manifests automatically to the physical classification schemes such as CMECS. Alternatives to in-situ sampling such as remote sensing classification offer a

Fig. 19 A self-organizing map classification of the Gulf of Mexico water for 2009. We trained, and applied 5 years of monthly data from 2005 to 2009 using a SOM

proxy for these classifiers, the environmental forcing functions, which shape and determine habitat suitability.

We have used machine learning to objectively provide a water classification of the US Gulf of Mexico using a suite of ocean data products. The unsupervised classification algorithm employed was a self-organizing map (SOM). We used SOM to reduce the dimensionality of the data set. We applied the SOM to identify the multivariate signatures of similar waters and to study the spatial and temporal trends of individual classes over a 5-year period. The input data employed was the sea surface temperature, the chlorophyll concentration, the sea surface salinity, the euphotic depth, and the difference between the sea-bottom and sea-surface temperatures. The output of the analysis is a comprehensive low-dimensional map of US Gulf of Mexico region. This may aid the decision making for the essential habitat zones for conservation and commercial purposes.

The result of the self-organizing map classification is summarized by color-coded maps as shown in Figs. 19 and 21. The SOM essentially group together similar regions as it picks out the characteristic signatures from the data. The similar colors represent the water classified by the method to be on the same class where as the different colors represent the different classification as depicted by the data signatures. Here, the dimension is essentially reduced from 5 to 1. For each SOM class we have computed the spatial extent in square km and how this varies with time. In the time series the annual and biannual signals are clearly evident. A good example of a clear annual cycle in the areal extent is Class 4 in the Eastern Gulf of Mexico. A good example of a clear biannual cycle in the areal extent is Class 61 in the Eastern Gulf of Mexico. Figure 20 shows examples using two SOM classes and the corresponding percentile distributions to the CMECS nomenclatures (Fig. 21).

The dynamics of the region dictates that the characteristics of water should change over the course of the year. In order to study such effects, we computed the areal extent of each of the SOM classes for each month over the 5-year period. By plotting the area of each SOM class as a function of time, we can clearly see the annual and bi-annual cycles in the GOM waters. Characteristic to their SOM class, the areas of the classes change depending upon the time of the year. This gives rise

Fig. 20 Webplots showing
the categorical variation of
data components for the SOM
classes 4 and 61. The
distributions of the variables
are represented by the 5th,
25th, 50th, 75th, and 95th
percentile values on the
corresponding axes. The
upper panel shows the
CMECS categories for SOM
class 4, associated with
waters that have warm sea
surface temperatures, are
euhaline, extremely clear,
oligotrophic and stratified.
The lower panel shows the
CMECS categories for the
SOM class 63, associated
with waters that are
temperate, mesohaline,
moderately clear,
mesotrophic, and mixed

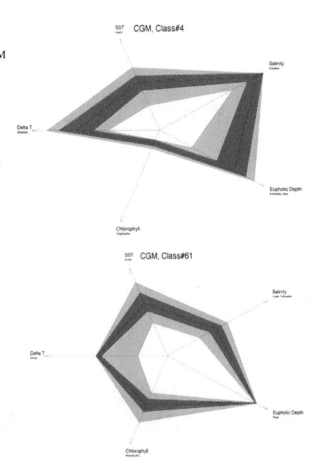

to a time series. The pattern repeats itself over the 5-year period. By using Fourier analysis we have identified the frequency distribution of the time series. Figure 22 shows two example SOM classes for the central Gulf of Mexico. We have computed the area of the SOM classes for each month for each class for all regions.

Class 61 lies in the near-shore region and has a clear bi-annual cycle. The areal extent of this class is greatest in April and October, and least in January. Similarly, SOM class 4, which lies in the offshore region, shows a clear annual cycle. The area of class 4 has a maximum during the summer months and has a minimum during the winter months.

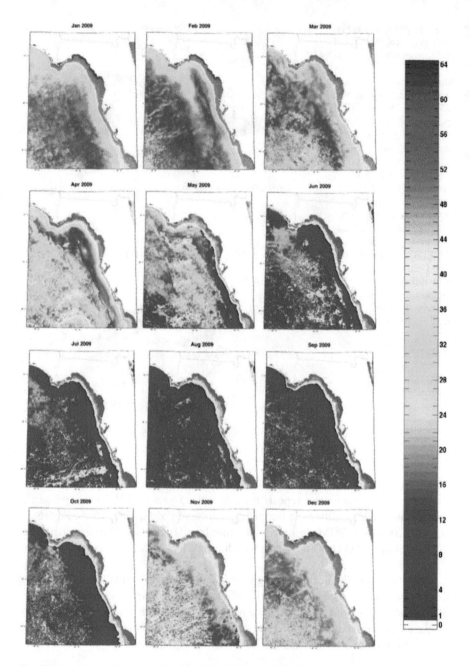

Fig. 21 A self-organizing map classification of Eastern Gulf of Mexico water for year 2009. We trained, and applied 5 years of monthly data from 2005 to 2009 using unsupervised artificial neural network technique called SOM. The color bar on the right shows the 64 classes corresponding to the colored region on the maps

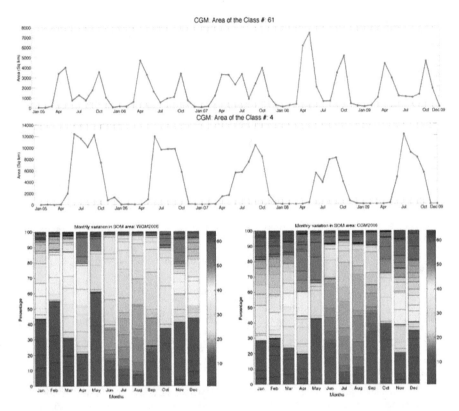

Fig. 22 The SOM classes show seasonal as well as annual trends. The area of SOM classes change depending upon the time of the year. For example, in the upper panels two classes are presented: Class 61 lies in the near-shore region and has a clear bi-annual cycle. The areal extent of this class is greatest in April and October, and least in January. Similarly, SOM class 4, which lies in the offshore region, shows a clear annual cycle. The area of class 4 has a maximum during the summer months and has a minimum during the winter months. The lower panel shows the monthly variation in the SOM class for 2006 for Western Gulf of Mexico (left) and Central Gulf of Mexico (right). The SOM classes in the range of 15–30 are dominant for the summer period (May–Sep)

Fish Catch and SOM Classes

The classes within the Self-organized Maps appear to correlate with fish counts. Figure 23 shows SEAMAP[8] fish count data for the western and central portions of the Gulf of Mexico in June 2006. The fish data are clearly clustered within classes 12–26. Those same classes dominate the SOM for June 2006. Further examination indicates that sea surface temperature is the driving factor for the dominant classes, with euphotic depth appearing to be the secondary driver.

---

[8]http://seamap.gsmfc.org/.

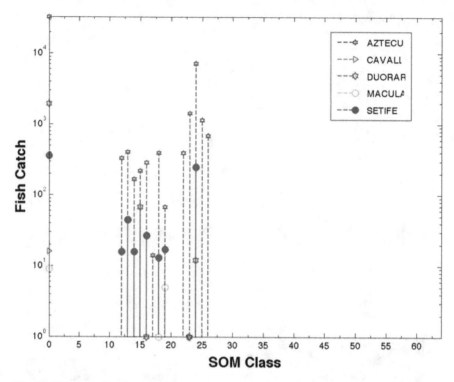

Fig. 23 Fish catch data from the Western Gulf of Mexico and Central Gulf of Mexico for June 2006 have been mapped to the SOM classes. The SOM classes in the range of 12–26 are the abundant classes and so is the fish catch

Some Likely Future Machine Learning Applications

Two recent advances are likely to open up a large number of new applications. First the improvement, size reduction and cost reduction of hyper-spectral imagery, and secondly, small embedded (credit card sized) GPU systems such as the NVIDIA Jetson TX1 with its 256 GPU cores.

Hyper-Spectral Imaging and Machine Learning for Real Time Embedded Processing and Decision Support

So what is Hyperspectral Imaging? The human eye perceives the color of visible light in three bands using the cones, the photoreceptor cells in the retina (Fig. 24). These three bands are red (centered on 564 nm), green (centered on 534 nm), and blue (centered on 420 nm). By contrast, instead of using just three broad bands, hyperspectral cameras divide the spectrum into a very large number of narrow

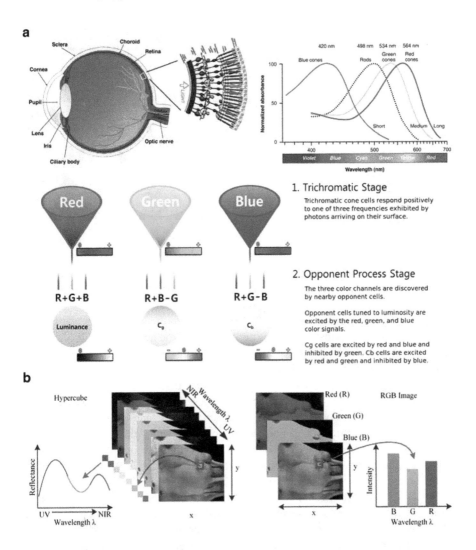

Fig. 24 Panel (**a**) Trichromatic cone cells in the eye respond to one of three wavelength ranges (RGB). These three bands are red (centered on 564 nm), green (centered on 534 nm), and blue (centered on 420 nm). Panel (**b**) shows a comparison between a hyperspectral-cube and RGB images. A hyper-cube is a three-dimensional dataset consisting of a two-dimensional image layers each for a different wavelength. So for each pixel in the image we have a multi-wavelength spectra (spectral signature). This is shown schematically in the lower left. On the right we see a conventional RGB color images with only three bands, images for red, green and blue wavelengths. The lower right shows an example 3 wavelength broad band spectra from a conventional RGB color image

Fig. 25 Hyperspectral cube

bands. Sometimes as many as two to four hundred bands are used to create a hyperspectral datacube (Fig. 25). This technique of dividing images into bands can extend beyond the visible, into both the infrared and thermal infrared, and into the ultraviolet.

Hyperspectral imaging systems are used around the world in a variety of medical, laboratory, industrial, agricultural, and airborne applications. To illustrate the broader significance, let us briefly review just some of these (Fig. 27). Hyperspectral imaging (HSI) is used in various medical applications, especially in disease diagnosis and image-guided surgery. The disease diagnosis applications (e.g., skin examination) naturally lend themselves to telemedicine applications for rural communities where the network connectivity can drastically improve rural community medical care. For each snapshot in time, HSI acquires a three-dimensional dataset called a hypercube (Fig. 25), with two spatial dimensions (just like a regular camera) and one spectral dimension, there is a separate collocated image/layer for each wavelength band (Fig. 24).

Spatially resolved spectral imaging obtained by HSI can provide diagnostic information about the tissue physiology, morphology, and composition. With the advantage of acquiring two-dimensional images across a wide range of electromagnetic spectrum, HSI has been applied to numerous areas, including archaeology and art conservation (Angeletti et al., 2005; Liang, 2012), vegetation and water resource control (Govender et al., 2007), food quality and safety control (Gowen et al., 2007; Feng and Sun, 2012), forensic medicine (Malkoff and Oliver, 2000; Edelman et al., 2012), crime scene detection (Muller et al., 2003), biomedicine (Afromowitz et al., 1988; Carrasco et al., 2003), agriculture, security and defense, thin films, etc.

Figure 27 shows some of the many HSI applications. For example: Using an airborne HSI, an invasive weed ("leafy spurge", Euphorbia esula) infestation could be clearly identified (Jay et al., 2010), and a weed coverage map generated (Fig. 27a). A study of seed germination (Nansen et al., 2015) using HSI showed that although viable and nonviable seeds appear identical to the human eye they can be clearly distinguished using full reflectance spectra (Fig. 27b). Analysis of wound healing (La Fontaine et al., 2014) (Fig. 27c). Mapping hydrological

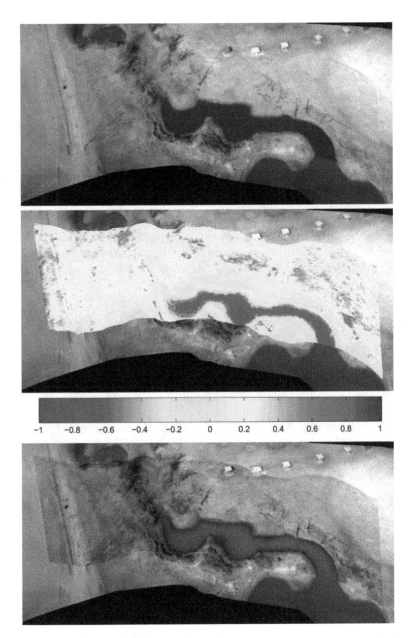

Fig. 26 Hyperspectral imaging of a rural landscape. Top image: sum of every spectral channel from the HS image, overlaid on top of the visible camera mosaic. Middle image: Normalized Difference Vegetation Index. Bottom image: pseudocolor from red, green, and blue channels (Ramirez, 2015)

formations (Fig. 27d). Fluorescent dye imaging (Fig. 27e). Examining the effect of surface pollution (Keith et al., 2009; Spangler et al., 2010) from leaking pipelines on vegetation (Fig. 27f). Checking food quality and fruit bruising (Fig. 27g). Classification of walnuts and shells (Fig. 27h). Automated analysis of cooked meats (Fig. 27i).

This diversity of examples demonstrates the general usefulness and applicability of HSI in a very broad range of contexts. In research, health, agriculture, industry, and more. We already saw in Sect. that combining the spectral signature in just seven wavelengths with machine learning was invaluable in uniquely identifying global dust sources with remarkable accuracy. So it can readily be seen that using more detailed hyper-spectral signatures with on-board embedded processing can provide incredibly powerful insights in a very compact package. Figure 26 shows an example of some hyperspectral imagery we obtained using our aerial vehicles (Ramirez, 2015). This approach is useful for many applications in smart agriculture, land surface classification, petrochemical surveying, disaster response (such as oil spills), etc. Let us take a closer look at the example of oil spill response.

Oil Spills

The National Academy of Sciences estimates 1.7–8.8 million tons of oil are released into global waters every year. More than 70% of this release is related to human activities. The effects of these spills include dead wildlife, oil covered marshlands and contaminated water (Fingas and Brown, 1997; Fingas, 2010; Liu et al., 2013; Cornwall, 2015). Spills of national significance (SONS), such as Deepwater Horizon (DWH), challenge response capabilities. In such large spills, *optimizing a coordinated response is a challenge.* There are always competing mission needs for aerial response resources such as helicopters and observer aircraft. Wildlife reconnaissance, oil observation overflights and targeting chemical dispersant application are a few examples. If we consider just one aspect, i.e. the spill itself, the challenges include both characterizing the continual temporal and spatial evolution of the spill extent, and the evolution of the oil itself as it weathers and emulsifies. Characterizing the oil spill can be made even more challenging due to the variable spill illumination and the weather. *Trained* observers are required, and their deployment needs can include a wide area, which is also challenging. Further, what is the optimal flight path(s) that should be used by the observers on each deployment to best meet the current needs and *anticipate the future evolution of the oil spill* to put in place any required preemptive measures or contingencies, such as shoreline pre-cleaning or protective boom deployment? During the DWH oil spill operational trajectory forecasting, maps of key areas for aerial observations to improve trajectory modeling were produced daily by NOAA for the overflight teams (Fig. 27).

The DWH oil spill and the associated impact monitoring was aided by extensive airborne and space-borne passive and active remote sensing (Fingas and Brown,

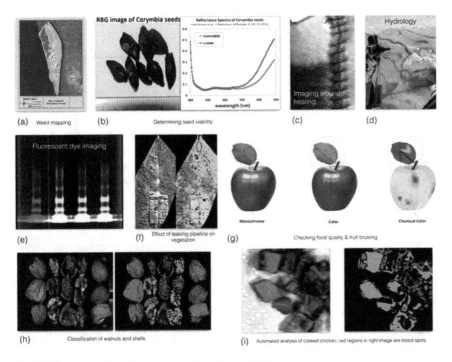

Fig. 27 Some examples of hyperspectral imaging applications

1997; Leifer et al., 2012; Liu et al., 2013; Fingas and Brown, 2014). A good review of these remote sensing activities is provided by (Leifer et al., 2012). During DWH, remote sensing was used to derive oil thickness (see Fig. 28) quantitatively for thick (>0.1 mm) slicks from AVIRIS (Airborne Visible/Infrared Imaging Spectrometer) that measured 224 contiguous spectral bands with wavelengths from 400 to 2,500 nanometers (nm) using a spectral library approach based on the shape and depth of near infrared spectral absorption features (Kokaly et al., 2013; Leifer et al., 2012). MODIS (Moderate Resolution Imaging Spectroradiometer) satellite, visible-spectrum broadband data of surface-slick modulation of sunglint reflection allowed extrapolation to the total slick. A multispectral expert system used a neural network approach to provide Rapid Response thickness class maps (Sveykovsky and Muskat, 2006; Svejkovsky et al., 2009).

Airborne and satellite synthetic aperture radar (SAR) provides synoptic data under all-sky conditions (Liu et al., 2011; Leifer et al., 2012); however, SAR generally cannot discriminate thick (>100 μm) oil slicks from thin sheens (to 0.1 μm). The UAVSAR's (Unmanned Aerial Vehicle SAR) significantly greater signal-to-noise ratio and finer spatial resolution allowed successful pattern discrimination related to a combination of oil slick thickness, fractional surface coverage, and emulsification.

Oil Code Color, Thickness, and Concentration Values

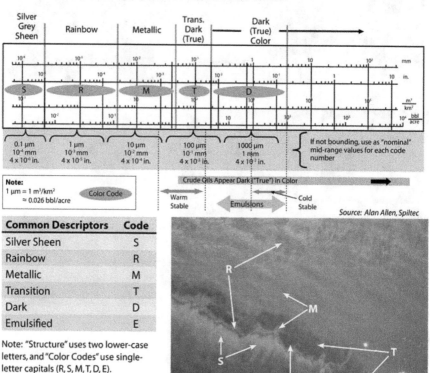

Common Descriptors

Common Descriptors	Code
Silver Sheen	S
Rainbow	R
Metallic	M
Transition	T
Dark	D
Emulsified	E

Note: "Structure" uses two lower-case letters, and "Color Codes" use single-letter capitals (R, S, M, T, D, E).

Fig. 28 Oil thickness chart and appearance from NOAA open water oil identification job aid

Further, in situ burning and smoke plumes were studied with AVIRIS and corroborated spaceborne CALIPSO (Cloud Aerosol Lidar and Infrared Pathfinder Satellite Observation) observations of combustion aerosols. CALIPSO and bathymetry lidar

data documented shallow subsurface oil, although ancillary data were required for confirmation.

Airborne hyperspectral, thermal infrared data have nighttime and overcast collection advantages, and were collected as well as MODIS thermal data. However, interpretation challenges and a lack of Rapid Response Products prevented significant use. Rapid Response Products were key to response utilization—*data needs are time critical*; thus, a high technological readiness level is vital to operational use of remote sensing products. The DWH oil spill experience demonstrated that development and operationalization of new near real-time spill response remote sensing tools must precede the next major oil spill (Leifer et al., 2012).

Cleanup of a SONS involve *multiple* skimmer ships, vessels collecting oil for in situ burning, and chemical dispersant operations. Typically these **slow** moving response ships are spread over a **large** area and guided by air support (e.g., helicopter), as the vessel bridge is too low to see the variation in thickness of the oil. Typically the manned air support will inform each ship of the location of recoverable oil ahead and then leave to overfly the next ship.

The cost of manned air support is significant, so each ship does *not* usually have dedicated continuous manned air support. In smaller spills, a report of oil location is given to the responding ship, these ships move slowly, so by the time they reach the location provided by manned air support, the oil has *moved*! For oil cleanup to be optimal (quickest) and effective (most oil recovered) the skimmer ships need to first focus on the regions of thickest oil. From the visual perspective of the ships crew, relatively close to the water and with a shallow viewing angle, it is not easy to know where the thickest oil is. Accurately discerning the gradations in black oil thickness is a challenging task.

Rich information on the thickness of the oil and the degree of weathering is contained in the detailed hyperspectral signature of the oil spill. This information can be utilized through the use of machine learning to provide real time response tools.

Summary

We have seen that machine learning has found many applications in remote sensing. These applications range from retrieval algorithms to bias correction, from code acceleration to detection of disease in crops, from classification of pelagic habitats to the rock type classification. As a broad subfield of artificial intelligence, machine learning is concerned with algorithms and techniques that allow computers to "learn." The major focus of machine learning is to extract information from data automatically by computational and statistical methods. Over the last decade there has been considerable progress in developing a machine learning methodology for a variety of Earth Science applications involving trace gases, retrievals, aerosol products, land surface products, vegetation indices, and most recently, ocean applications.

References

7 million premature deaths annually linked to air pollution. http://www.who.int/mediacentre/news/releases/2014/air-pollution/en/. Accessed 29 Aug 2016

Afromowitz MA, Callis JB, Heimbach DM, Desoto LA, Norton MK (1988) Multispectral imaging of burn wounds - a new clinical instrument for evaluating burn depth. IEEE Trans Biomed Eng 35(10):842–850. https://doi.org/10.1109/10.7291. <GotoISI>://WOS:A1988Q389400009

Andrews CP, Ratner PH, Ehler BR, Brooks EG, Pollock BH, Ramirez DA, Jacobs RL (2013). The mountain cedar model in clinical trials of seasonal allergic rhinoconjunctivitis. Ann Allergy Asthma Immunol 111(1):9–13

Angeletti C, Harvey NR, Khomitch V, Fischer AH, Levenson RM, Rimm DL: Detection of malignancy in cytology specimens using spectral-spatial analysis. Lab Invest 85(12):1555–1564. https://doi.org/10.1038/labinvest.3700357. <GotoISI>://WOS:000233372200011

Arizmendi C, Sanchez J, Ramos N, Ramos G (1993) Time series predictions with neural nets: application to airborne pollen forecasting. Int J Biometeorol 37(3):139–144

Aurin M, Mannino A (2012) A database for developing global ocean color algorithms for colored dissolved organic material, CDOM spectral slope, and dissolved organic carbon, paper presented at Ocean Optics XXI, The Oceanography Society, Glasgow, UK

Bishop CM (1995) Neural networks for pattern recognition. Oxford University Press, Oxford (1995)

Bouet C, Cautenet G, Bergametti G, Marticorena B, Todd MC, Washington R (2005) Sensitivity of desert dust emissions to model horizontal grid spacing during the Bodele Dust Experiment 2005. Atmos Environ 50, 377–380 (2012)

Breiman L (1984) Classification and regression trees. The Wadsworth statistics/probability series. Wadsworth International Group, Belmont, CA

Breiman L (2001) Random forests. Mach Learn 45(1):5–32 (2001). Times Cited: 3621

Britton J, Pavord I, Richards K, Knox A, Wisniewski A, Wahedna I, Kinnear W, Tattersfield A, Weiss S (1994) Factors influencing the occurrence of airway hyperreactivity in the general population: the importance of atopy and airway calibre. Eur Respir J 7(5):881–887

Brown ME, Lary DJ, Vrieling A, Stathakis D, Mussa H (2008) Neural networks as a tool for constructing continuous ndvi time series from AVHRR and MODIS. Int J Remote Sens 29(24):7141–7158

Carrasco O, Gomez R, Chainani A, Roper W (2003) Hyperspectral imaging applied to medical diagnoses and food safety. In: Proceedings of the society of photo-optical instrumentation engineers (SPIE), vol 5097, pp. 215–221. https://doi.org/10.1117/12.502589 <GotoISI>://WOS:000185395500023

Castellano-Méndez M, Aira M, Iglesias I, Jato V, González-Manteiga W (2005) Artificial neural networks as a useful tool to predict the risk level of betula pollen in the air. Int J Biometeorol 49(5):310–316

Committee FGD, et al. (2012) Coastal and marine ecological classification standard. Publication# FGDC-STD-018-2012

Cornwall W (2015) Deepwater horizon: after the oil. Science 348(6230):22–29. https://doi.org/10.1126/science.348.6230.22. http://www.sciencemag.org/content/348/6230/22.short

Cortes C, Vapnik V (1995) Support-vector networks. Mach Learn 20(3):273–297. Times Cited: 3429

Csépe Z, Makra L, Voukantsis D, Matyasovszky I, Tusnády G, Karatzas K, Thibaudon M (2014) Predicting daily ragweed pollen concentrations using computational intelligence techniques over two heavily polluted areas in Europe. Sci Total Environ 476:542–552

D'amato G, Spieksma FTM (1991) Allergenic pollen in Europe. Grana 30(1):67–70

Demuth HB, Beale MH, De Jess O, Hagan MT (2014) Neural network design, 2nd edn. Martin Hagan, Stillwater (2014)

Domingos P (2015) The master algorithm: how the quest for the ultimate learning machine will remake our world. Basic Books, New York

Edelman GJ, Gaston E, van Leeuwen TG, Cullen PJ, Aalders MCG (2012) Hyperspectral imaging for non-contact analysis of forensic traces. Forensic Sci Int 223(1–3):28–39. https://doi.org/10. 1016/j.forsciint.2012.09.012. <GotoISI>://WOS:000311432100021

Ernst P, Ghezzo H, Becklake M (2002) Risk factors for bronchial hyperresponsiveness in late childhood and early adolescence. Eur Respir J 20(3):635–639

Esch RE, Hartsell CJ, Crenshaw R, Jacobson RS (2001) Common allergenic pollens, fungi, animals, and arthropods. Clin Rev Allergy Immunol 21(2):261–292

Feng YZ, Sun DW (2012) Application of hyperspectral imaging in food safety inspection and control: a review. Crit Rev Food Sci Nutr 52(11):1039–1058. https://doi.org/10.1080/ 10408398.2011.651542. <GotoISI>://WOS:000306740000007

Fingas M (2010) Oil spill science and technology. Gulf Professional Publishing, Houston (2010)

Fingas MF, Brown CE (1997) Review of oil spill remote sensing. Spill Sci Technol Bull 4(4):199–208. http://dx.doi.org/10.1016/S1353-2561(98)00023-1. http://www.sciencedirect. com/science/article/pii/S1353256198000231. The Second International Symposium on Oil Spills

Fingas M, Brown C (2014) Review of oil spill remote sensing. Mar Pollut Bull 83(1):9–23. http:// dx.doi.org/10.1016/j.marpolbul.2014.03.059. http://www.sciencedirect.com/science/article/pii/ S0025326X14002021

Friedman J, Hastie T, Tibshirani R (2001) The elements of statistical learning. Springer Series in Statistics, vol 1. Springer, Berlin

Genuer R, Poggi JM, Tuleau-Malot C (2010) Variable selection using random forests. Pattern Recogn Lett 31(14):2225–2236

Govender M, Chetty K, Bulcock H (2007) A review of hyperspectral remote sensing and its application in vegetation and water resource studies. Water SA 33(2):145–151. <GotoISI>:// WOS:000246960100001

Gowen AA, O'Donnell CP, Cullen PJ, Downey G, Frias JM (2007) Hyperspectral imaging - an emerging process analytical tool for food quality and safety control. Trends Food Sci Technol 18(12):590–598. https://doi.org/10.1016/j.tifs.2007.06.001. <GotoISI>://WOS: 000251485800001

Haykin SS (1994) Neural networks: a comprehensive foundation. Macmillan, New York. 93028092 Simon Haykin. ill. ; 26 cm. Includes bibliographical references (pp 635–690) and index

Haykin SS (1999) Neural networks: a comprehensive foundation, 2nd edn. Prentice Hall, Upper Saddle River, NJ. 98007011 Simon Haykin. ill. ; 25 cm. Includes bibliographical references (pp 796–836) and index

Haykin SS (2001) Kalman filtering and neural networks. Adaptive and learning systems for signal processing, communications, and control. Wiley, New York. 2001049240 edited by Simon Haykin. ill. ; 24 cm. "A Wiley Interscience publication." Includes bibliographical references and index

Haykin SS (2007) New directions in statistical signal processing : from systems to brain. Neural information processing series. MIT Press, Cambridge, MA. 2005056210 GBA671791 013536699 (OCoLC)ocm62302576 (OCoLC)62302576 edited by Simon Haykin ...[et al.]. ill. ; 26 cm. Includes bibliographical references (p. [465]–508) and index. Modeling the mind : from circuits to systems/Suzanna Becker – Empirical statistics and stochastic models for visual signals/David Mumford – The machine cocktail party problem/Simon Haykin, Zhe Chen – Sensor adaptive signal processing of biological nanotubes (ion channels) at macroscopic and nano scales/Vikram Krishnamurthy – Spin diffusion : a new perspective in magnetic resonance imaging/Timothy R. Field – What makes a dynamical system computa- tionally powerful?/Robert Legenstein, Wolfgang Maass – A variational principle for graphical models/Martin J. Wainwright, Michael I. Jordan – Modeling large dynamical systems with dynamical consistent neural networks/Hans-Georg Zimmermann ...[et al.] – Diversity in com- munication : from source coding to wireless networks/Suhas N. Diggavi – Designing patterns for easy recognition : information transmission with low-density parity-check codes/Frank R. Kschischang, Masoud Ardakani – Turbo processing/Claude Berrou, Charlotte Langlais, Fabrice Seguin – Blind signal processing based on data geometric properties/Konstantinos Diamantaras – Game-theoretic learning/Geoffrey J. Gordon – Learning observable operator models via the efficient sharpening algorithm/Herbert Jaeger ...[et al.]

Ho TK (1998) The random subspace method for constructing decision forests. IEEE Trans Pattern Anal Mach Intell 20(8):832–844

Howard LE, Levetin E (2014) Ambrosia pollen in tulsa, oklahoma: aerobiology, trends, and forecasting model development. Ann Allergy Asthma Immunol 113(6):641–646

Jay SC, Lawrence RL, Repasky KS, Rew LJ (2010) IEEE: detection of leafy spurge using hyperspectral-spatial-temporal imagery. In: 2010 IEEE international geoscience and remote sensing symposium, pp 4374–4376. https://doi.org/10.1109/igarss.2010.5652580. <GotoISI>://WOS: 000287933804134

Kasprzyk I (2008) Non-native ambrosia pollen in the atmosphere of rzeszów (se poland); evaluation of the effect of weather conditions on daily concentrations and starting dates of the pollen season. Int J Biometeorol 52(5):341–351

Keith CJ, Repasky KS, Lawrence RL, Jay SC, Carlsten JL (2009) Monitoring effects of a controlled subsurface carbon dioxide release on vegetation using a hyperspectral imager. Int J Greenhouse Gas Control 3(5):626–632. https://doi.org/10.1016/j.ijggc.2009.03.003. <GotoISI>://WOS: 000268949300013

Kinney PL (2008) Climate change, air quality, and human health. Am J Prev Med 35(5):459–467

Kokaly RF, Couvillion BR, Holloway JM, Roberts DA, Ustin SL, Peterson SH, Khanna S, Piazza SC (2013) Spectroscopic remote sensing of the distribution and persistence of oil from the deepwater horizon spill in barataria bay marshes. Remote Sens Environ 129:210–230. http://dx.doi.org/10.1016/j.rse.2012.10.028. http://www.sciencedirect.com/science/article/pii/S0034425712004166

Koren I, Kaufman YJ, Washington R, Todd MC, Rudich Y, Martins JV, Rosenfeld D (2006) The Bodele depression: a single spot in the Sahara that provides most of the mineral dust to the Amazon forest. Environ Res Lett 1(1):011001

Kotsiantis SB, Zaharakis I, Pintelas P (2007) Supervised machine learning: A review of classification techniques

Laaidi M, Laaidi K, Besancenot JP, Thibaudon M (2003) Ragweed in france: an invasive plant and its allergenic pollen. Ann Allergy Asthma Immunol 91(2):195–201

La Fontaine J, Lavery L, Zuzak K (2014) The use of hyperspectral imaging (HSI) in wound healing. Proc SPIE 8979. https://doi.org/897903 10.1117/12.2041841. <GotoISI>://WOS: 000336032300001

Lary D, Aulov, O (2008) Space-based measurements of HCl: intercomparison and historical context. J Geophys Res: Atmos 113(D15)

Lary D, Müller M, Mussa H (2003) Using neural networks to describe tracer correlations. Atmos Chem Phys Discuss 3(6):5711–5724

Lary D, Waugh D, Douglass A, Stolarski R, Newman P, Mussa H (2007) Variations in stratospheric inorganic chlorine between 1991 and 2006. Geophys Res Lett 34(21)

Lary DJ, Remer LA, MacNeill D, Roscoe B, Paradise S (2009) Machine learning and bias correction of MODIS aerosol optical depth. IEEE Geosci Remote Sens Lett 6(4):694–698

Lary DJ, Faruque FS, Malakar N, Moore A, Roscoe B, Adams ZL, Eggelston Y (2014) Estimating the global abundance of ground level presence of particulate matter (pm2. 5). Geospat Health 8(3):611–630

Lary DJ, Alavi AH, Gandomi AH, Walker AL (2016) Machine learning in geosciences and remote sensing. Geosci Front 7(1):3–10

Leifer I, Lehr WJ, Simecek-Beatty D, Bradley E, Clark R, Dennison P, Hu Y, Matheson S, Jones CE, Holt B, Reif M, Roberts DA, Svejkovsky J, Swayze G, Wozencraft J (2012) State of the art satellite and airborne marine oil spill remote sensing: application to the BP deepwater horizon oil spill. Remote Sens Environ 124:185–209. http://dx.doi.org/10.1016/j.rse.2012.03. 024. http://www.sciencedirect.com/science/article/pii/S0034425712001563

Lellis-Dibble KA, McGlynn K, Bigford TE (2008) Estuarine fish and shellfish species in US commercial and recreational fisheries: economic value as an incentive to protect and restore estuarine habitat. National Oceanic and Atmospheric Administration, National Marine Fisheries Service, Office of Habitat Conservation, Habitat Protection Division (2008)

Lewis WH, Vinay P, Zenger VE (1983) Airborne and allergenic pollen of North America. Johns Hopkins University Press, Baltimore

Liang H (2012) Advances in multispectral and hyperspectral imaging for archaeology and art conservation. Appl Phys A-Mater Sci Process 106(2):309–323

Liu P, Li X, Qu JJ, Wang W, Zhao C, Pichel W (2011) Oil spill detection with fully polarimetric {UAVSAR} data. Mar Pollut Bull 62(12):2611–2618. http://dx.doi.org/10.1016/j.marpolbul. 2011.09.036. http://www.sciencedirect.com/science/article/pii/S0025326X11005248

Liu Y, MacFadyen A, Ji Z, Weisberg R (2013) Monitoring and modeling the deepwater horizon oil spill: a record breaking enterprise. Geophysical monograph series. Wiley, Hoboken

Low RB, Bielory L, Qureshi AI, Dunn V, Stuhlmiller DF, Dickey DA (2006) The relation of stroke admissions to recent weather, airborne allergens, air pollution, seasons, upper respiratory infections, and asthma incidence, september 11, 2001, and day of the week. Stroke 37(4): 951–957

Malkoff DB, Oliver WR (2000) Hyperspectral imaging applied to forensic medicine. Progr Biomed Opt 1:108–116. <GotoISI>://WOS:000086469300013

Matheson EM, Player MS, Mainous AG, King DE, Everett CJ (2008) The association between hay fever and stroke in a cohort of middle aged and elderly adults. J Am Board Fam Med 21(3):179–183

McCulloch W, Pitts W (1943) Bull Math Biophys 5:115. https://doi.org/10.1007/BF02478259

Muller MG, Valdez TA, Georgakoudi I, Backman V, Fuentes C, Kabani S, Laver N, Wang ZM, Boone CW, Dasari RR, Shapshay SM, Feld MS (2003) Spectroscopic detection and evaluation of morphologic and biochemical changes in early human oral carcinoma. Cancer 97(7):1681–1692. https://doi.org/10.1002/cncr.11255. <GotoISI>://WOS:000181816600012

Nansen C, Zhao G, Dakin N, Zhao C, Turner SR (2015) Using hyperspectral imaging to determine germination of native Australian plant seeds. J Photochem Photobiol B-Biol 145:19–24. https://doi.org/10.1016/j.jphotobiol.2015.02.015. <GotoISI>://WOS:000352679200003

Nowosad J (2016) Spatiotemporal models for predicting high pollen concentration level of corylus. Int J Biometeorol. Springer 60(6):843–855

Osowski S, Garanty K (2007) Forecasting of the daily meteorological pollution using wavelets and support vector machine. Eng Appl Artif Intell 20(6):745–755

Oswalt ML, Marshall GD (2008) Ragweed as an example of worldwide allergen expansion. Allergy Asthma Clin Immunol 4(3):130

Puc M (2012) Artificial neural network model of the relationship between betula pollen and meteorological factors in szczecin (Poland). Int J Biometeorol 56(2):395–401

Ramirez DA (1984) The natural history of mountain cedar pollinosis. J Allergy Clin Immunol 73(1):88–93

Ramirez JP, Lary DJ, Gans NR (2015) Low-altitude terrestrial spectroscopy from a pushbroom sensor. J Field Robot 1–16. https://doi.org/10.1002/rob.21624

Rodríguez-Rajo F, Astray G, Ferreiro-Lage J, Aira M, Jato-Rodriguez M, Mejuto JC (2010) Evaluation of atmospheric poaceae pollen concentration using a neural network applied to a coastal atlantic climate region. Neural Netw 23(3):419–425

Safavian SR, Landgrebe D (1991) A survey of decision tree classifier methodology. IEEE Trans Syst Man Cybern 21(3):660–674. https://doi.org/10.1109/21.97458

Sánchez-Mesa J, Galán C, Martínez-Heras J, Hervás-Martínez C (2002) The use of a neural network to forecast daily grass pollen concentration in a mediterranean region: the southern part of the iberian peninsula. Clin Exp Allergy 32(11):1606–1612

Sassen K (2008) Boreal tree pollen sensed by polarization lidar: depolarizing biogenic chaff. Geophys Res Lett 35(18):L18810

Spangler LH, Dobeck LM, Repasky KS, Nehrir AR, Humphries SD, Barr JL, Keith CJ, Shaw JA, Rouse JH, Cunningham AB, Benson SM, Oldenburg CM, Lewicki JL, Wells AW, Diehl JR, Strazisar BR, Fessenden JE, Rahn TA, Amonette JE, Barr JL, Pickles WL, Jacobson JD, Silver EA, Male EJ, Rauch HW, Gullickson KS, Trautz R, Kharaka Y, Birkholzer J, Wielopolski L (2010) A shallow subsurface controlled release facility in bozeman, montana, usa, for testing near surface CO_2 detection techniques and transport models. Environ Earth Sci 60(2):227–239. https://doi.org/10.1007/s12665-009-0400-2. <GotoISI>://WOS:000276637000002

Stark PC, Ryan LM, McDonald JL, Burge HA (1997) Using meteorologic data to predict daily ragweed pollen levels. Aerobiologia 13(3):177–184

Svejkovsky J, Muskat J, Mullin J (2009) Adding a multispectral aerial system to the oil spill response arsenal. Sea Technol 50(8):17–22

Sveykovsky J, Muskat S (2006) Real-time detection of oil slick thickness patterns with a portable multispectral sensor. Tech. rep. (2006)

Todd MC, Washington R, Martins JV, Dubovik O, Lizcano G, M'Bainayel S, Engelstaedter S (2007) Mineral dust emission from the bodele depression, northern chad, during bodex 2005. J Geophys Res-Atmos 112(D6):D06207

Tränkle E, Mielke B (1994) Simulation and analysis of pollen coronas. Appl Opt 33(21): 4552–4562

Vapnik VN (1982) Estimation of dependences based on empirical data. Springer series in statistics. Springer, New York

Vapnik VN (1995) The nature of statistical learning theory. Springer, New York

Vapnik VN (2000) The nature of statistical learning theory. Statistics for engineering and information science, 2nd edn. Springer, New York

Vapnik VN (2006) Estimation of dependences based on empirical data ; empirical inference science : afterword of 2006. Information science and statistics, 2nd enl. edn. Springer, New York, NY

Voukantsis D, Niska H, Karatzas K, Riga M, Damialis A, Vokou D (2010) Forecasting daily pollen concentrations using data-driven modeling methods in Thessaloniki, Greece. Atmos Environ 44(39):5101–5111

Washington R, Todd MC (2005) Atmospheric controls on mineral dust emission from the bodele depression, chad: the role of the low level jet. Geophys Res Lett 32(17):4543

Washington R, Todd MC, Engelstaedter S, Mbainayel S, Mitchell F (2006) Dust and the low-level circulation over the bodele depression, chad: observations from bodex 2005. J Geophys Res-Atmos 111(D3):D03201

Washington R, Todd MC, Lizcano G, Tegen I, Flamant C, Koren I, Ginoux P, Engelstaedter S, Bristow CS, Zender CS, Goudie AS, Warren A, Prospero JM (2006) Links between topography, wind, deflation, lakes and dust: the case of the Bodele depression, chad. Geophys Res Lett 33(9):L09401

Wayne P, Foster S, Connolly J, Bazzaz F, Epstein P (2002) Production of allergenic pollen by ragweed (ambrosia artemisiifolia l.) is increased in CO2-enriched atmospheres. Ann Allergy Asthma Immunol 88(3):279–282

Zhao F, Elkelish A, Durner J, Lindermayr C, Winkler JB, Ruëff, F., Behrendt, H., Traidl-Hoffmann, C., Holzinger, A., Kofler, W., et al.: Common ragweed (*Ambrosia artemisiifolia* L.): allergenicity and molecular characterization of pollen after plant exposure to elevated NO2. Plant Cell Environ 39(1):147–164

5

Mapping Floods and Assessing Flood Vulnerability for Disaster Decision-Making: A Case Study of Remote Sensing Application

Bessie Schwarz, Gabriel Pestre, Beth Tellman, Jonathan Sullivan, Catherine Kuhn, Richa Mahtta, Bhartendu Pandey, and Laura Hammett

Introduction

While environmental and social threats to society changes faster than in recent centuries, there is more of a need for faster, globally scalable and locally relevant risk information from developing Banks and the countries they serve. Big Data can range from gigabytes (call details records), to terabytes (satellite data), to petabytes (web traffic), with each magnitude requiring unique algorithms to extract the signal from the noise. This chapter explores how one type of sensor data—satellite imagery—can be made more useful through the development of an application that leverages Cloud Computing—Google Earth Engine—to turn data into insight for decision-makers on the ground.

The huge availability of and increasing computing power for satellite imagery is essentially meaningless for practical development without demand-driven applications to turn the information into insights and ensure those insights are locally relevant. We need an "application revolution" in step with the data revolution

---

B. Schwarz (✉) • B. Tellman • J. Sullivan • C. Kuhn • R. Mahtta • L. Hammett
Cloud to Street, USA
e-mail: bessie@cloudtostreet.info; Beth@cloudtostreet.info; jonsull@umich.edu; ckuhn@uw.edu; richa.mahtta3@gmail.com; lmhammett@gmail.com

G. Pestre
The Data Pop Alliance, New York, NY, USA
e-mail: gpestre@datapopalliance.org

B. Pandey
Yale University, New Haven, CT, USA
e-mail: bhartendu.pandey@yale.edu

(UN Data Revolution Group 2014). To take one use case, each year millions of people and billions of dollars' worth of assets around the world are affected by floods, which cause more economic, social, and humanitarian losses worldwide than any other type of hazard (UNISDR 2015). Yet the communities most at risk from environmental and societal change, especially those in the developing world, often lack the information they need to help understand, prepare for, and respond to these threats.

For one example, this chapter describes a satellite imagery based rapid mapping tool that distills massive datasets about flood vulnerability into useful risk information for Senegal. This approach—which combines machine learning, remote sensing, and census data to predict the socio-physical vulnerability to floods and dynamically delivers that information to decision-makers through a web-platform—demonstrates the potential for operationalizing remote sensing data to better understand and address information gaps about flood risk in Senegal and around the world.

More broadly, this case study helps reveal opportunities for integrating Earth Observation with other types of data. Combining data from mobile phones, social media, and more with satellite imagery and official statistics can add a near-real-time layer to remotely sensed land and disaster detection, creating deforestation alerts and disaster response tools for instance. Analysis of Call Detail Records (CDR) data provides a picture of population distribution at a finer temporal scale, showing where people are at the moment a remotely sensed flood hits an area. This can be critical in areas where high rates of seasonal migration or daily commuting. CDR data can also estimate how and where affected populations migrate after a disaster (Wilson et al. 2016; Lu et al. 2016). Pairing news scraping technologies with Earth Observation data via cloud computing platforms like Earth Engine make it possible to provide more accurate spatial detection that respond to input from observers and decision-makers on the ground (see Fig. 3). These new data mining techniques make it possible to update flood maps more quickly or learn of the existence of floods that are not mentioned in international disaster declarations or available in current databases like The Dartmouth Flood Observatory.

Why Senegal? An Example of the Critical Information Gap in Disaster Management

In Senegal, flood risk is constantly shifting in due to changing climate and urban settlement. As in much of the Sahel, Senegal has experienced a history of highly uncertain climatic conditions, varying between cycles of drought to eras of frequent and severe flooding. After several very dry decades (1968–1997), the average rainfall increased 35% between 2000 and 2005 (Nicholson 2005). In addition to changing climate, Senegal has undergone significant land use change, triggered by extreme drought events in the 1970s, 80s, and 90s and forcing rural populations into urban areas (Goldsmith et al. 2004). The peak urbanization rate of Senegal's capital, Dakar, was estimated around 7–8%, and 44% of Senegalese currently live in urban areas (Mbow et al. 2008).

Despite the high degree of vulnerability and rapidly changing risk in Senegal, there is an inadequate risk data and vulnerability modeling for Senegal. Without this critical flood risk information, communities, governments, donors, and others in development lack clear guidance on how to prioritize resources, which limits their ability to make effective infrastructure investments, devise emergency plans, or provide relief assistance when needs arise (Hellmuth et al. 2011; Mitchell et al. 2010).

Socio-Physical Vulnerability to Flooding to Senegal

Cloud to Street's approach to vulnerability combines new big data analysis tools with rapid assessment disaster science to fill information gaps about exposure and vulnerability to flooding in Senegal. With the support of Agence Française de Développement, social and physical vulnerability models were developed for Senegal and combined to determine the country's exposure to flooding and estimate its social vulnerability to future hazard.

Flood detection: The method first estimates flood exposure by creating a historic inventory of major floods in Senegal. A list of past flood events in Senegal was assembled from a number of publicly available information sources (see Fig. 1).

Machine learning hydrology: Using the inventory of past floods as training data, a machine learning model was developed in Google Earth Engine to predict which parts of the country and population are at risk from future extreme flood for five priority watersheds. The floodplains cover 34% of the country. The Saint-Louis region, in the Senegal River Valley, was the primary testing ground for customizing the algorithm, where the authors designed and assessed four machine learning approaches on 11 flood conditioning factors. The model's high accuracy

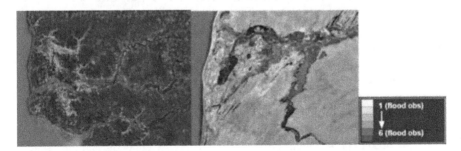

Fig. 1 Number of times an area (pixel) flooded from 2003–2015 in Senegal using the DFO algorithm at 250 m per pixel resolution (*left*: Ziguinchor, Senegal; *right*: Saint-Louis, Senegal and Senegal River)

rate for predicting training data demonstrate that machine learning algorithms can successfully predict floods using remote sensing (58–98%, depending on the watershed).

Social vulnerability to flooding: Identifying the social conditions that make one community more likely to experience loss from a disaster—loss of life, loss of livelihood, lack of recovery—is critical to understanding the threat of and resilience to flooding in Senegal. Experts at Cloud to Street conducted a literature review and PCA-based factor analysis to assess social vulnerability for Senegal, using anonymized data from Senegal's 2013 census data, obtained through a partnership with Data-Pop Alliance and the Agence Nationale de la Statistique et de la Démographie du Sénégal.

Results: In the five priority watersheds, the method predicts a floodplain of 5596 km^2. Of this area, 30% is high-risk zone where over 97,000 people live. Additionally, approximately five million people live in the 30 *arrondissements* that have very high social vulnerability profiles compared to other arrondissements. Five underlying dimensions that drive vulnerability in Senegal: (1) a lack of basic informational resources, (2) old age, (3) disabilities, (4) being disconnected from dense hubs and (5) population increase from internal migration. These five factors explain ~69% of the variation in the selected census variables.

Combined socio-physical vulnerability of Senegal: Several of the arrondissements identified as having high biophysical risk were also found to have high or very high social vulnerability (see Fig. 2). These preliminary results show promise for cheaper and faster ways to gather flood information critical to disaster management and risk reduction, although a complete nation-wide assessment of the biophysical risk profile would be necessary to yield insights into the combined socio-physical vulnerability, since this assessment only considered five priority watersheds. Because the method relies on Earth Observing satellites, new information can be added in each flood event to retrain machine learning algorithms on the fly and improve prediction accuracies as the model responds to new training data.

Using Big Data Information in Disaster Management

The flood risk information from this assessment can provide valuable insights for decision-makers at all stages of the disaster management cycle—from disaster risk reduction and event prediction to response and recovery (see Fig. 3). A flood vulnerability assessment map is a valuable hotspot analysis, indicating areas of high need to prioritize with preparedness funding and resources. Development Banks who fund infrastructure builds, social programs and other projects to can better target investments and protect assets. National and State governments in flood prone areas who need to protect their citizens, their citizens' livelihoods, and their state's economy over the long term, can prepare areas that are most at risk and communities that are most likely to experience loss when hit by hazards. Search and rescue agencies and humanitarian NGOs, aimed at minimize human casualties and reduce

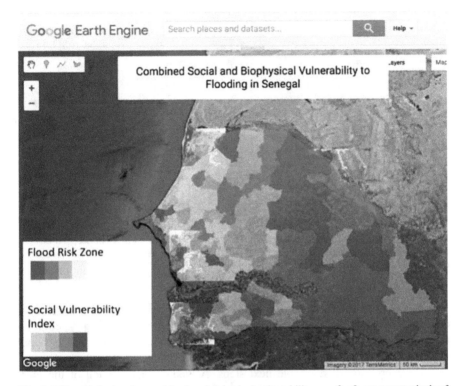

Fig. 2 This map depicts the combined socio-physical vulnerability map for five test watersheds of Senegal and a mock demonstration of the Senegal Flood Risk Dash-board

immediate and long term suffering from flooding, can use a real-time version for this assessment to mobilize support to locations in need when help is most critical aka in the 24 h immediately after a disaster.

The Future of This Approach Globally and Locally as a Practical Tool

The science and technology outlined here is a baseline assessment with more research to be done to fully operationalize it for global use. This test model produces floodplain predictions that range from 30–250 m in spatial scale because it was trained on publicly available NASA satellites, but the same machine learning model could produce results at finer resolution if appropriate imagery were available for floods in the country. Increasing the spatial and temporal resolution of the flood model using private data (5–1 m resolution, with a daily or weekly return period, as is possible with the Planet satellite fleet), expanding the inventory of mapped floods for Senegal (but scraping news and using crowdsourcing), predicting the floodplain for

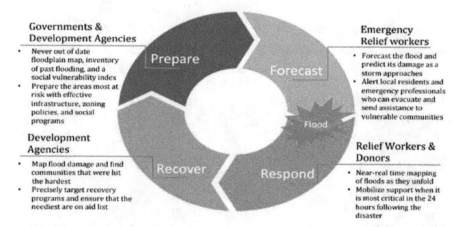

Governments & Development Agencies
- Never out of date floodplain map, inventory of past flooding, and a social vulnerability index
- Prepare the areas most at risk with effective infrastructure, zoning policies, and social programs

Development Agencies
- Map flood damage and find communities that were hit the hardest
- Precisely target recovery programs and ensure that the neediest are on aid list

Emergency Relief workers
- Forecast the flood and predict its damage as a storm approaches
- Alert local residents and emergency professionals who can evacuate and send assistance to vulnerable communities

Relief Workers & Donors
- Near-real time mapping of floods as they unfold
- Mobilize support when it is most critical in the 24 hours following the disaster

Fig. 3 The flood risk information can provide valuable insights for a variety of decision-makers at all stages of the disaster management cycle—from disaster risk reduction and event prediction to response and recovery

the entire country (by automating optimized machine learning models in the cloud), adding flood depth and economic damage predictions (by developing function to extract depth from optical and radar images), and using predictive analytics and private data in social vulnerability analysis (using cellphone data released by Orange to researchers) are all reachable next steps. A web-enabled version of the map results could be set to stream satellite imagery from public and private sensors, and accepting crowdsourced contributions in near real time, so that the vulnerability analysis for Senegal can be updated each time satellites beam new imagery to earth with the mere refresh of a browser page, which can be essential in developing countries.

Remote sensing based tools for development need to scale globally so that virtually everyone can take advantage of the data revolution, especially as risk changes faster than ever, but this information has to also always be taken in context with other situational information from the ground when used for local or even larger scale decision-making. Applications of remote sensing tools for development need to be designed with the flexibility and humanity needed to protect the dignity of the community being analyzed and engage to them in analysis from the start. Future stages of the vulnerability assessment described in this chapter can engage local communities to verify the location of historic flooding, "ground-truth" the social vulnerability assessment indicators, and add fidelity to the machine learning based physical flooding analysis. For instance, digital crowdsourcing applications and on the ground participatory science could increase the accuracy and richness of the science and help build local resilience through risk awareness and better national disaster management.

Conclusion

Satellite imagery now produces a wealth of data that can be fused with other data and processing techniques such as CDRs and news scraping. Yet the wealth of data captured from above the earth and analyzed in the cloud does not always trickle down into information that can be used on the ground to promote development and increase resilience. There is therefore a need, and great potential, for applications that integrate new data sources and techniques with existing Earth Observation science to turn data into useful information to analyze threats to development such as flood hazards and enable managers on the ground to respond.

References

Goldsmith PD, Gunjal K, Ndarishikanye B (2004) Rural-urban migration and agricultural productivity: the case of Senegal. Agric Econ 31(1):33–45. https://doi.org/10.1111/j.1574-0862.2004.tb00220.x

Hellmuth ME, Mason SJ, Vaughan C, van Aalst MK, Choularton R (eds) (2011) A better climate for disaster risk management. International Research Institute for Climate and Society (IRI), Columbia University, New York, NY. https://iri.columbia.edu/wp-content/uploads/2013/07/CSP3_Final.pdf

Lu X, Wrathall DJ, Sundsøy PR, Nadiruzzaman M, Wetter E, Iqbal A, Qureshi T, Tatem AJ, Canright GS, Engø-Monsen K, Bengtsson L (2016) Detecting climate adaptation with mobile network data in Bangladesh: anomalies in communication, mobility and consumption patterns during cyclone Mahasen. Clim Chang 138(3–4):505–519. https://doi.org/10.1007/s10584-016-1753-7

Luo T, Maddocks A, Ward P, Winsemius H (2015) World's 15 countries with the most people exposed to river floods. World Resources Institute. http://www.wri.org/blog/2015/03/world%E2%80%99s-15-countries-most-people-exposed-river-floods

Mbow C, Diop A, Diaw AT, Niang CI (2008) Urban sprawl development and flooding at Yeumbeul suburb (Dakar-Senegal). Afr J Environ Sci Technol 2(4):75–88. http://www.ajol.info/index.php/ajest/article/view/135466

Mitchell T, Ibrahim M, Harris K, Hedger M, Polack E, Ahmed A, Hall N, Hawrylyshyn K, Nightingale K, Onyango M, Adow M, Sajjad Mohammed S (2010) Climate smart disaster risk management. Strengthening Climate Resilience, Institute of Development Studies (IDS), Brighton, UK. http://www.drrprojects.net/drrp/default/download/doc_document.file.a5bb8f7dbadd042b.435344524d2e706466.pdf

Nicholson S (2005) On the question of the "recovery" of the rains in the West African Sahel. J Arid Environ 63(3):615–641. https://doi.org/10.1016/j.jaridenv.2005.03.004

Schwarz B, Tellman B, Sullivan J, Kuhn C, Mahtta R, Pandey B, Hammett L, Pestre G (2017) Socio-physical vulnerability to flooding in Senegal: an exploratory analysis with new data & Google Earth Engine (AFD technical reports no. 25). Agence Française de Développement, Paris, France. http://librairie.afd.fr/nt25-va-vunerability-flooding-senegal/

UNISDR (2015) The human cost of weather related disasters 1995–2015. United Nations Office for Disaster Risk Reduction. https://www.unisdr.org/we/inform/publications/46796

United Nations Data Revolution Group (2014) Data revolution report. United Nations Secretary-General's Independent Expert Advisory Group on a Data Revolution for Sustainable Development. http://www.undatarevolution.org/report/

Wilson R, zu Erbach-Schoenberg E, Albert M, Power D, Tudge S, Gonzalez M, Guthrie S, Chamberlain H, Brooks C, Hughes C, Pitonakova L, Buckee C, Lu X, Wetter E, Tatem A, Bengtsson L (2016) Rapid and near real-time assessments of population displacement using mobile phone data following disasters: the 2015 Nepal earthquake. PLOS Currents Disasters. https://doi.org/10.1371/currents.dis.d073fbece328e4c39087bc086d694b5c

6

New Generation Platforms for Exploration of Crowdsourced Geo-Data

Maria Antonia Brovelli, Marco Minghini, and Giorgio Zamboni

Abstract This chapter addresses two recent topics in the field of geo-information, the former more technological and the latter more scientific. On one side, there is an emerging trend of visualizing data and their changes in space and time through multidimensional geospatial clients and/or virtual globes. In the most advanced cases, these are not simply plain viewers but also allow analysis of the data by acting as "multidimensional intelligent geo-viewers". On the other side, citizen science is providing a great momentum to the possibility of lay people taking part in scientific development. It is a new, citizen-centred paradigm which, in most cases, takes advantage of the individual and collective augmented capability of sensing the surrounding world through the sensors that we wear. The "citizen sensors" will consciously contribute to this development, either through volunteered geographic information or by being themselves an unconscious part of the data analytics, which makes use of geo-crowdsourced data to extract information in order to create a higher level understanding of natural and manmade phenomena. This chapter seeks to outline the Web technological solutions for visualizing and analyzing such data, through a summary of the current state of the art and the original applications developed by the authors.

Introduction

Since the beginning of the new century a dramatic change has characterized the evolution of GeoWeb, defined as the collection of services and data supporting the use of geographic information over the Internet (Scharl and Tochtermann 2007). While the delivery of geospatial data over the Web began in the early 1990s (Putz 1994), it was not until 2005 that—following the success of Web 2.0 (O'Reilly 2005)—a new wave of geospatial Web applications has marked the birth of the so-called GeoWeb 2.0 (Maguire 2007). During the last two decades Web mapping applications have gradually evolved from being mostly static, poorly interactive

M.A. Brovelli (✉) • M. Minghini • G. Zamboni
Department of Civil and Environmental Engineering, Politecnico di Milano, Piazza Leonardo da Vinci 32, 20133, Milano, Italy
e-mail: maria.brovelli@polimi.it; marco.minghini@polimi.it; giorgio.zamboni@polimi.it

and with limited functionality (few data accessible and few functions available on them) to dynamic, interactive and resembling the traditional desktop counterparts (Plewe 2007). Many factors have contributed to this success such as the increase of computational power of computers and broadband Internet communication, the introduction of technologies such as AJAX (Asynchronous JavaScript and XML) and mapping APIs (Application Programming Interfaces) as well as the OGC (Open Geospatial Consortium) standards for Web mapping interoperability (Peng and Tsou 2003).

Almost 10 years ago a new boost to the exploitation of Web-served geospatial data was given by the birth of virtual globes, and more in general multidimensional clients offering spatio-temporal visualization of geographic contents. Beyond their successful impact on the general public (Butler 2006), virtual globes introduced powerful ways not only to visualize but also to explore, analyse and process the heterogeneous and complex data generated by scientific disciplines (Blower et al. 2007).

Another big shift in the history of GIS began at almost the same time. Summarized by the term Volunteered Geographic Information or VGI (Goodchild 2007), a new era started where users have become active providers of geospatial contents. Social networking, digital photography, online mapping and mobile devices are currently fruitful sources of VGI, with OpenStreetMap (Arsanjani et al. 2015) being the most popular example of a collaborative spatial data collection project. Citizen science, outlining volunteers' participation in the development of scientific activities (Cohn 2008), relies in many cases on VGI. This citizen-centred paradigm takes advantage of the individual augmented capability of sensing the world, which thanks to the sensors we wear can also produce vast quantities of unconsciously geo-crowdsourced data—termed by Harvey (2013) as CGI or Contributed GI (as opposed to VGI). One such example being the amount of data which is daily collected by smart phone users almost without any user control.

How mature is the bridge between virtual globes and VGI? The purpose of this chapter is to contribute to the discussion by investigating two innovative ways virtual globes can be exploited, namely as: (a) collaboration platforms for visualizing, exploring and collecting VGI; and (b) multidimensional intelligent geo-viewers providing advanced spatio-temporal visualization of citizen-sensed data. These topics are deepened through the description of two specific applications which have been recently (early 2016) developed by the authors.

The remainder of the chapter is structured as follows. Section on "Virtual Globes as the New Generation Visualization Platforms" provides a background on virtual globes which outlines their historical development, primary features and main software implementations. Next, a review is provided on virtual globe-based applications focused on the exploration of crowdsourced geo-data. The two original applications developed by the authors are then presented in section on "Applications": a collaborative platform for managing citizen-sensed information and an intelligent geo-viewer of telecommunication data. Section on "Conclusions" concludes with the summary of the work's contributions and directions for future research.

Virtual Globes as the New Generation Visualization Platforms

As mentioned above, virtual globes have ushered in a historical change in the way geospatial information is delivered over the Internet. This era was identified by Plewe (2007) as the fourth generation of Web Mapping applications, coming to the forefront after three different stages of 2D applications featuring an increasing level of interaction and usability. Virtual globes are software providing a 3D representation of the Earth and allowing users to explore it in a virtual environment while streaming satellite imagery, Digital Elevation Models (DEMs) and other geographic data from the Internet (Schultz et al. 2008). Unlike traditional GIS they offer users a "magic carpet ride" above the Earth surface. Users can also interact with the information represented on or above the surface, as well as manage geographic layers and optionally add their own data to create 3D mash-ups.

Key factors which enabled the birth of virtual globes include the development of high-performance graphics cards for 3D visualization (mainly driven by the video game industry), the evolution of Internet bandwidth, the increased availability of high-resolution satellite imagery (Yu and Gong 2012) and the advancements in the field of global spatial grids (Sahr et al. 2003). The success of virtual globes must be certainly attributed to Google Earth,[1] which since its release in 2005 has gained an astonishing popularity among the general public. Goodchild reasonably claimed that "Just as the PC democratized computing, so systems like Google Earth will democratize GIS" (Butler 2006). In response to Google Earth ESRI released ArcGIS Explorer[2] and Microsoft launched its Bing Maps Platform[3] (previously Microsoft Virtual Earth), while an open source alternative—World Wind[4] released by NASA—has been available since 2003.

Many features of virtual globes contributed to their popularity. Virtual globes are multi-scale, multipurpose and multidimensional, i.e. they can even provide a 4D visualization if time component is considered (Brovelli et al. 2013). Virtual globes offer high context capability while at the same time providing ease and intuitiveness of use. In turn this translates into a richer and more realistic user experience. What is more, the chance of visualizing, locating and interacting with different kinds of geospatial contents within a single 3D virtual environment has opened many possibilities for using virtual globes in a number of geo-related disciplines, including scientific ones (Brovelli and Zamboni 2012).

In addition to those mentioned above, several other virtual globe platforms have been released over the last decade. Examples are ESRI ArcGlobe,[5] SkylineGlobe,[6]

---

[1]http://www.google.com/earth

[2]http://www.esri.com/software/arcgis/explorer

[3]http://www.microsoft.com/maps

[4]http://worldwind.arc.nasa.gov/java

[5]http://www.esri.com/software/arcgis/extensions/3danalyst

[6]http://www.skylinesoft.com

CitySurf Globe,[7] SuperMap GIS,[8] EarthBrowser,[9] ossimPlanet,[10] QGIS Globe,[11] gvSIG3D[12] (based on NASA World Wind), osgEarth,[13] Marble,[14] Norkart Virtual Globe[15] and Earth3D[16]. Standards like HTML5 and WebGL[17] have recently made it possible to render interactive 3D graphics within Web browsers without the need of specific proprietary applications, run-time environments (e.g. Java Virtual Machine) or plugins. A new wave of WebGL-based virtual globes has thus emerged including Cesium,[18] WebGL Earth,[19] WebGL Globe,[20] OpenWebGlobe[21] (Loesch et al. 2012) and the Web-based version of World Wind[22]. It is out of the scope of this chapter to provide a rigorous and detailed description of the characteristics and differences between the virtual globes listed above. However, some useful classifications of virtual globes according to a number of criteria (platform-dependence, license type, type of application, geographic layers available by default, functionalities and freedom of customization) are provided by Brovelli et al. (2014). Keysers (2015) proposes a categorization of virtual globes into platforms and visualization applications: open and closed source; public and restricted access.

An emerging concept related to virtual globes is Digital Earth (DE). It was envisioned by Gore (1998) as a multi-resolution, 3D representation of the planet allowing scientists, policy-makers, students and all societal sectors to manage georeferenced information and navigate them in space and time (Brovelli et al. 2015). Though DE and virtual globes are often used as synonyms, according to Harvey (2009) DE refers to the experiences and knowledge people have, and continue to make of the world, while virtual globes represent the most viable term for discussions of DE software environment. In other words, virtual globes are not DE, but rather a point of access to DE. This is why in the growing literature on DE virtual globes are also named geobrowsers (see for example Goodchild 2008;

---

[7] http://www.citysurf.com.tr

[8] http://www.supermap.com

[9] http://www.earthbrowser.com

[10] https://trac.osgeo.org/ossim/wiki/OssimPlanet

[11] http://hub.qgis.org/wiki/quantum-gis/Globe_Plugin

[12] https://redmine.gvsig.net/redmine/projects/gvsig-3d/wiki

[13] http://osgearth.org

[14] https://edu.kde.org/marble

[15] http://www.virtual-globe.info

[16] http://www.earth3d.org

[17] https://www.khronos.org/webgl

[18] https://cesiumjs.org

[19] http://www.webglearth.org

[20] https://github.com/dataarts/webgl-globe

[21] http://www.openwebglobe.org

[22] http://webworldwind.org

Annoni et al. 2011; Craglia et al. 2012), thus underlining their correspondence to traditional Web browsers in searching, retrieving and organizing information, whose nature is however spatial (Jones 2007).

Virtual Globes for the Exploration of Crowdsourced Geo-Data

Despite having democratized access to geospatial information, due to their primarily commercial orientation (driven by Google Earth), virtual globes were initially considered poorly suited for scientific purposes (Craglia et al. 2008). Some years later Goodchild et al. (2012) identified some limits of the existing virtual globes compared to Gore's vision, e.g. the inability to explore the Earth both backward and forward in time, the limited set of available data, and the user interaction which often resembled an access to a searchable catalogue rather than a real exploitation. Nonetheless, the capabilities of virtual globes described in section on "Virtual Globes as the New Generation Visualization Platforms" have made them increasingly used in a variety of scientific disciplines. At the time of writing (early 2016), examples include geology (Blenkinsop 2012; Martínez-Graña et al. 2013; Zhu et al. 2014), ecology (Guralnick et al. 2007), history and cultural heritage (Brovelli et al. 2013; Valentini et al. 2014), cartographic heritage (Brovelli et al. 2012; Gede et al. 2013), natural disasters and disaster management (Webley 2011; Tomaszewski 2011), environmental analyses and modelling (Chien and Tan 2011; Brovelli and Zamboni 2012), weather forecast (Smith and Lakshmanan 2011), instruction and education (Lindner-Fally and Zwartjes 2012; Bodzin et al. 2014), health (Stensgaard et al. 2009), landscape planning (Schroth et al. 2011) and urban studies (Nebiker et al. 2010). User functions typically include data visualization, collection, exploration, integration, validation, communication and dissemination, modelling, and decision support (Yu and Gong 2012).

In line with the main content of this chapter, the following discussion provides an overview of virtual globe applications focused on geo-data volunteered or crowdsourced by users. It is worth noting that VGI has been recognized as a key opportunity for the development of DE (Craglia et al. 2008; Annoni et al. 2011) and that even OpenStreetMap has been shown to be well-positioned within DE's goals (Mooney and Corcoran 2014).

Elvidge and Tuttle (2008) predict the power of virtual globes in democratizing content contributions by users. Examples related to Google Earth are described such as the Maplandia gazetteer and the geotagged photos from Panoramio. Using virtual globes as pure visualization platforms to display geotagged contents from media sharing sites is still a common practice. As an example, Kisilevich et al. (2010) exploit Google Earth to produce interactive visualizations of tourist sites based on geotagged photos from Panoramio and Flickr. Similarly, Vu et al. (2015) make use of Google Earth to visualize Flickr photos for exploring tourists' travel behaviour. Both examples lack time filtering of data.

Cooper et al. (2010) draft a questionnaire to explore the perceptions held by geographical information professionals from Africa on the use of VGI in virtual globes. Results reveal a number of pros and cons, these latter including the fact that digital divide (Norris 2001) is still a strong barrier to the democratization promoted by VGI and virtual globes. Gede et al. (2013) created a virtual museum where a WebGL-based virtual globe environment allows users to navigate digitized paper globes. Upon registration, the users can add/edit both data and descriptions, thus adding a participative nature to the collection. Wu et al. (2010) describe a 3D urban planning interoperable environment based on GeoGlobe, a virtual globe prototype developed by the authors. Both urban planning designers and the public can visualize, manipulate and discuss urban planning projects. Participation happens through the upload of urban models as well as online tools such as forums, e-mails and instant messaging. Some spatial analysis tools are also available within the platform (e.g. to compute distances and calculate/simulate sunlight and sunshade). Beltran et al. (2013) present the VisioMIMEXT application built on top of NASA World Wind which allows the user to spatially search, visualize and retrieve georeferenced media resources from diverse social networks (Twitter, OpenStreetMap, YouTube, Flickr, Geonames and Wikipedia). Time filtering of data is planned as a future extension. The authors stress the role of virtual globes as data fusion tools for both experts and the public.

Virtual globes are also recognized as useful Web-based tools for collaborative mapping. The collaborative geospatial environment described by Nebiker et al. (2007) allows integrating observations from real-world mobile geo sensors (e.g. GPS logs and UAV video streams) into a three-dimensional viewer (i3D) developed by the authors. Additional communication channels between users such as chat, video or voice are also available. Similarly, Yovcheva et al. (2010) describe the development of a geo-collaborative web-based virtual globe prototype, based on Google Earth and designed for distributed teams of environmental researchers working with spatio-temporal data.

Virtual globes are also used in remote sensing as platforms for collecting, comparing and validating land use/land cover (LULC) data. The main example is Geo-Wiki[23] (Fritz et al. 2009; Fritz et al. 2012), an online Google Earth-based crowdsourcing platform where volunteers exploit high-resolution imagery to perform spatial validation of existing LULC maps. A recent paper (See et al. 2015) offers an overview of the main features of the platform together with examples on how the crowdsourced data collected through Geo-Wiki has been used to improve information on LULC. A similar system is VIEW-IT (Virtual Interpretation of Earth Web-Interface Tool), a crowdsourcing platform based on Google Earth, which allows users to collect LULC data from high-resolution imagery (Clark and Aide 2011). Another tool designed for experts to assess LULC change and validate global LULC maps is developed by Bastin et al. (2013). The platform is mainly 2D but makes also use of Google Earth to increase context capability. Depending on the area, users can visualize images at different dates using a time slider.

---

[23]http://geo-wiki.org

Applications

In contrast to the picture emerging from the literature review outlined in section on "Virtual Globes as the New Generation Visualization Platforms", two virtual globe-based applications are presented in the following which constitute promising examples of new generation 3D/4D platforms to explore volunteered and crowdsourced data. The first, described in section on "PoliCrowd", provides a general-purpose collaboration platform for managing observations collected from mobile devices. A multidimensional viewer of telecommunication data, based on a high-level platform for scientific data analysis, is then presented in section on "Telecommunication Data Viewer".

PoliCrowd

The first application, named PoliCrowd[24] (current version 2.0), implements a multidimensional and multi-thematic collaboration platform based on NASA World Wind virtual globe. PoliCrowd is a Web-based system able to: (a) allow citizens to perform field surveys using common mobile devices, compiling and submitting ad hoc questionnaires together with multimedia contents (images, videos, etc.), which are registered through the device sensors and georeferenced through the device positioning services (e.g. the GPS); (b) store and manage field-captured data into a spatial database and provide Web publication using standard protocols; and (c) provide a Web-based collaboration platform offering not just a 3D/4D visualization interface, but also an overall framework allowing users to manage data and create customized maps.

PoliCrowd is fully built upon existing open source software and is in turn released as open source. The two main software products exploited for the development are the Open Data Kit[25] (ODK) suite for managing citizen-sensed data, and the World Wind virtual globe on which the platform is built. After a detailed description of these components, the system architecture and its main functionalities are presented in the following.

Open Data Kit (ODK)

The ODK suite is composed of three main modules providing complementary functions to perform and manage data collection, i.e. ODK Build, ODK Aggregate

---

[24] http://geomobile.como.polimi.it/policrowd2.0

[25] https://opendatakit.org

Fig. 1 Example of ODK Collect form to report cultural POIs

and ODK Collect. ODK Build[26] enables design of the form (i.e. the questionnaire) that users will complete in the field. It is an HTML5 Web application offering a drag-and-drop interface, which works well for designing simple forms. When complex forms need to be designed, XLSForm[27] can be used instead. The form is then loaded into ODK Aggregate, which can be deployed on Google App Engine (enabling users to quickly get running without facing the complexities of setting up their own scalable Web service) or locally on a Tomcat[28] server backed with a MySQL[29] or PostgreSQL[30] database. ODK Aggregate is the server-side component of the ODK suite and it is also responsible for user administration. It provides blank forms to, and accepts compiled forms from ODK Collect, which is the client-side component, which runs on Android mobile devices. ODK Collect allows users to perform point-wise surveys including different data types (text, number, location, multimedia, barcodes) and works well without a network connection. An example of ODK Collect form to report cultural Points of Interest (POIs) is shown in Fig. 1.

NASA World Wind Java SDK

Used by space agencies (e.g. the ESA—European Space Agency), local and national government agencies (e.g. the US FAA—Federal Aviation Administration), commercial industry and scientific communities, NASA World Wind is based on open standards and can be configured to load and visualize the most popular spatial data formats. It is conceived to work with large amounts of data and information, all

---

[26]http://build.opendatakit.org

[27]http://xlsform.org

[28]http://tomcat.apache.org

[29]http://www.mysql.com

[30]http://www.postgresql.org

of which can be stored on local disks or remote data servers. Many satellite images are available by default including BlueMarble (BMNG 500 m), i3-Landsat (15 m), USGS Orthophoto (1 m US only), USGS Urban Area Orthophoto (0.5–0.25 m selected US cities), and MS Virtual Earth Aerial Imagery, as well as some Digital Terrain Models (DTMs) including SRTM30Plus (30 arc-sec, 900 m), SRTM3 v2 v4.1 (3 arc-sec, 90 m), ASTER (30 m), and USGS NED (30 m, 10 m US only). Both images and DTMs are dynamically served by NASA[31] and USGS[32] WMSs (Web Mapping Services). Moreover, it is possible to access any OGC compliant WMS server and specific servers providing DEMs that can be superimposed onto the geoid model implemented within the platform. Finally, it is possible to locate on the globe, and in its surrounding space, both 2D objects (lines, polygons, markers, callouts and multimedia viewers) and 3D objects built up from geometric primitives (parallelepipeds, spheres, extruded polygons, etc.). The platform is suitable for any geospatial application thanks to the possibility of controlling the structure, quality and accuracy of the surface horizontal and vertical components (by varying the textures and the DTMs, respectively) and the full three-dimensional component of the virtual globe (the 2D and 3D objects that can be created for specific application scenarios).

The openness of World Wind code ensures a complete control of both the customization and the extensions required for creating the 3D model and all functionalities needed to interact with the model itself. Moreover, a platform-independent viewer can be implemented which is directly executable on different operating systems and potentially accessible by a simple Web browser. As mentioned in section on "Virtual Globes as the New Generation Visualization Platforms", currently there are two versions of World Wind: the first one, simply named NASA World Wind, is written in Java and uses the JOGL library[33]; the second one, named Web World Wind, is written in HTML5 and JavaScript (JS) and makes use of WebGL. Both platforms are available as a Software Development Kit (SDK) and provide the infrastructure to create spatial applications based on virtual globe technology. Currently the Java version provides a number of advanced features, which are not yet implemented in the JS version, and for this reason we chose to use the NASA World Wind Java SDK. Being written in Java, it is both platform-independent and usable as both a desktop application or accessible by simple Web browsers as an applet or via Java Web Start (JWS) technology.

System Architecture

A technical overview of the PoliCrowd 2.0 platform is shown in Fig. 2. The platform consists of two main components: the PoliCrowd client and the PoliCrowd server.

---

[31] http://www.nasa.gov

[32] https://www.usgs.gov

[33] http://jogamp.org/jogl/www

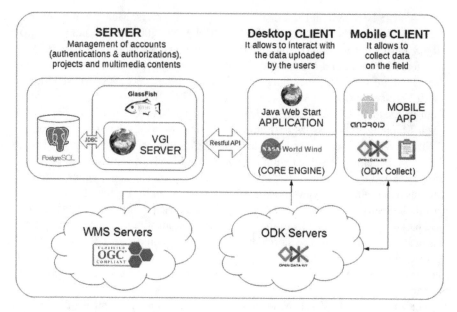

Fig. 2 System architecture of the PoliCrowd 2.0 platform

The former, available as a JWS application, is based on World Wind and developed by taking advantage of its SDK. It provides a 3D visualization of the available geospatial contents and interacts with the PoliCrowd server through a Restful Web Service. The server is deployed on the GlassFish[34] J2EE (Java 2 Enterprise Edition) application server backed with a PostgreSQL relational database to store all of the data related to the platform (system data and data uploaded by the users). The PoliCrowd client can connect to and retrieve data from any WMS and ODK server, thus allowing visualization and interaction on the globe with contents created through ODK Collect mobile clients.

Overview of Functionalities

After a first prototypal stage in which PoliCrowd was customized for tourism and cultural topics (Brovelli et al. 2014), the current version, PoliCrowd 2.0, represents its natural evolution with the purpose of being a completely general-purpose and multi-thematic social platform. Using a simple and intuitive graphical user interface (GUI) users can create, populate, customize and share projects, which refer to one or more specific topics integrated and displayed simultaneously within the same scenario.

[34]https://glassfish.java.net

Fig. 3 3D visualization of markers corresponding to POIs collected by users through ODK Collect

The 3D viewer allows users to interact with two kinds of geo-data. First, the application can connect to any WMS compliant server. A first level of information is provided by several base layers such as OSM, the aerial orthophotos and satellite imagery, and the thematic maps made available as standard geoservices by the geoportals of institutional administrations (e.g. INSPIRE and national geoportals) or by any other dedicated WMS server specifically chosen for the customization of the platform. A second level of information is provided by the crowdsourced data collected by users through the ODK Collect mobile app. More in detail, the platform has an innovative capability to connect with any ODK server and publish the related point-wise data through a user-customizable marker-based representation (see Fig. 3).

The markers are then open to collaborative contribution: any PoliCrowd user can add comments and upload multimedia contents (images, video and audio files) in order to enrich and share the information on each POI. When clicking on a POI marker, the related details collected through ODK Collect are displayed (see Fig. 4) and a dedicated section provides access to the textual and multimedia contents (comments, images, audios and videos). All multimedia contents are managed by the PoliCrowd client: images are displayed into a picture viewer whereas, for videos and audios, the application relies on a customizable external player depending on the user's preference.

Fig. 4 Examples of data visualization of POIs collected by citizens: bike parking in Osaka city, Japan (*left*) and panoramic view on hiking trails in Como city, Italy (*right*)

POIs visualization is fully customizable thanks to a suitable layer management interface. The attributes to be displayed can be filtered out and the marker icons can be customized either by choosing them from a default collection or by uploading them manually. The application can manage different projects populated with different WMS and ODK layers. All the projects created by users are publicly accessible by design in order to promote community participation. Finally, the introduction of the fourth dimension (time) provides a more in-depth navigation through the content. The time bar enables temporal filtering of all the POIs just by picking a given date or setting a range of dates. The same capability is also available for multimedia contents, which can be independently filtered according to the date of upload or (if provided) the actual date they were created.

The PoliCrowd functionalities briefly described so far will be presented with abundance of details in the following.

PoliCrowd can display a wide variety of data types since it can be connected to any WMS and ODK server available on the Internet. More in detail the ODK API, which is integrated into the virtual globe Java code, enables the PoliCrowd client to retrieve all the information required to connect to any ODK Aggregate server, i.e. the list of available forms and the corresponding submitted data, which are then visualized on the platform. From a logical perspective, the connection to an ODK server is treated in the same way as to a standard WMS server: the user has to select the type of server, enter its URL and then pick up the layer(s) of interest (see Fig. 5). The only difference is that, while WMS servers provide raster map layers, ODK servers provide marker layers.

Any server (both WMS and ODK) a user connects to is saved into a persistent library which is available to all PoliCrowd users in order to facilitate its reuse. Once users have created customized collections of layers, they can save their projects in a catalogue which is available for the whole community. The users' credentials, the list of available servers and projects, and the projects themselves (i.e. the available layers and their on/off status, the location and the camera orientation of the point of view over the globe) are stored in the PostgreSQL database on the PoliCrowd server.

Fig. 5 Form to transparently connect to any WMS or ODK server available on the Internet (*left*) and form to select the layer(s) published by an ODK server (*right*)

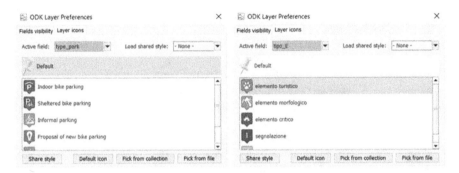

Fig. 6 Marker icons for bike parking (*left*) and for environmental POIs (*right*)

Consistent with its multi-thematic nature, PoliCrowd allows to dynamically customize marker icons on the platform client-side according to the specific project the user is working on. Unlike WMS layers, where the graphic layout is predefined on server side (WMS layer style), the PoliCrowd client provides an intuitive graphical interface to define the styles for the point-wise marker layers derived from ODK servers. When adding an ODK layer to a project, users can select the point attribute to use for styling layer markers among those available, e.g. the type of point. Automatically, as it happens for the typical styling tools in GIS, all the values assumed by the markers for the selected attribute are displayed and for each of them, users can either pick up an icon from a rich library of icons available within the platform, or they can provide an icon by manually uploading the file (see Fig. 6). In the same way of servers and projects, styles are also shareable, so that users adding the same layer in their projects can save time by opting for a ready-made icon set defined by another user.

The social nature of PoliCrowd allows any user to add comments and multimedia (images, audios and videos) to any ODK layer. This is possible thanks to a built-in interface on the client-side available when clicking on a marker. Besides showing all the information related to that marker, the interface displays the available multimedia contents and allows users to add new ones (see Fig. 7). PoliCrowd integrates all these functionalities in the same client-side graphics environment, thus making visualization and data upload easy and intuitive. Moreover, a dedicated

Fig. 7 New multimedia contents can be added by users and managed by the platform administrator using a dedicated Web application

Web application based on the Java Server Faces[35] technology was designed and implemented to facilitate the remote administration of the platform contents, enabling all the CRUD (create, read, update and delete) basic functions on the database.

PoliCrowd 2.0 introduces the fourth dimension in data visualization. A temporal slider available on the bottom side of the interface (see Fig. 7) allows time filtering of both the ODK marker layers and their multimedia contents. While markers are filtered according to their submission date (i.e. the date they were created from ODK Collect), multimedia contents can be filtered either by the date they were uploaded on the platform or, if available, by the actual date the content was created (i.e. the picture was taken, the video was shot, the audio was recorded). Users can optionally add the date of creation, together with a brief description of the multimedia content, as a metadata during the upload. Time filtering capability makes PoliCrowd particularly suited for historical projects and, more in general, projects where temporal information plays a key role.

Any user (authenticated or anonymous) can access, view and explore all the projects created within the platform and automatically shared with the community. Each project is maintained by its creator (owner). This means that the metadata (project name, project description, etc.), the list of layers used (both WMS and ODK) and the marker styles can be only edited by the project owner. Authenticated users can create and share new projects by adding WMS and ODK layers, uploading new POIs through ODK Collect, and adding comments and multimedia contents to both their own POIs and to preexisting POIs uploaded by other users. The authorization policies ensure that only the owners of the context (projects, POIs, comments, and multimedia contents) can perform editing operations on it.

---

[35] http://www.oracle.com/technetwork/java/javaee/javaserverfaces-139869.html

Telecommunication Data Viewer

Geo-crowdsourced data can be typically visualized and analyzed in two different ways. In the first case, each contribution is treated as a single, self-consistent information and managed as such (e.g. the POIs collected and published in the PoliCrowd application). In the second case, all contributions are pre-processed in order to derive thematic maps describing a widespread phenomenon in a global view. Using this second methodology, starting from data collected by the "citizen sensors" a spatial distributions of the related variables of interest (social, economic, environmental, etc.) is modelled through a geographical grid where each cell value is derived through interpolation functions, sampling and statistical aggregations (Brovelli and Zamboni 2012).

The netCDF Data Format

One of the most widely used storage formats to handle grid-based data in science is netCDF (Network Common Data Form), a machine-independent data format supporting the creation, access and sharing of array-oriented data (Rew et al. 2010). NetCDF is used for storing n-dimensional rectangular structures containing many kinds of scientific data in various fields of observation (climate, ocean, atmospheric, pollution, chemistry, medical imaging, etc.). It is a self-describing, portable, scalable, appendable, shareable and archivable format. In details, a netCDF file includes information about the data it contains (self-describing); it can be accessed by computers with different ways of storing integers, characters and floating-point numbers (portable); a small subset of a large dataset may be accessed efficiently (scalable); data may be appended to a properly structured netCDF file without copying the dataset or redefining its structure (appendable); one writer and multiple readers may simultaneously access the same netCDF file (shareable); access to all earlier forms of netCDF data will be supported by current and future versions of the software (archivable).

NetCDF is not only a machine-independent data format for representing data, but also a set of software interfaces and libraries which support the creation, access and sharing of scientific data. The netCDF software implements an abstract data type, which means that all operations to access and manipulate data in a netCDF dataset must use only the set of functions provided by the interface. The representation of the data is hidden from applications using the interface, so that the way data are stored could be changed without affecting existing programs. The physical representation of netCDF data is designed to be independent of the computer on which the data were written.

EST-WA: A Multidimensional Data Viewer

In the open source domain, a number of applications and libraries are available to visualize and manage netCDF data. Almost all of them rely on a bidimensional approach to visualize and analyse the data, while the third-dimension information (e.g. the height of the data above the ground level) is indirectly represented by means of contour lines, slide sections, graduated colour ramps or numerical values acquired by querying the geometries shown in the 2D visualization. The fourth dimension information (i.e. the time) is instead managed by providing a static "snapshot" of the data at a specific selected point in time. On the opposite EST-WA[36] (Environment Space and Time Web Analyzer) is a tool based on netCDF format able to visualize spatio-temporal variable distributions (e.g. sea temperature or soil permeability at different depths, air pollution, temperature, and humidity at different heights) in a multidimensional interface where both space (the 3D scenario) and time are simultaneously shown (Brovelli and Zamboni 2014).

In its general architecture written in Java, EST-WA uses the netCDF Java library, which extends the netCDF core data model and adds functionality implementing Unidata's Common Data Model (CDM), an abstract data model designed for scientific datasets. The Common Data Model has three layers, which are built on top of each other to add increasingly richer semantics. The "data access layer" (also known as the syntactic layer) handles data reading and writing; the "coordinate system layer" identifies the coordinates of the data arrays; and the "scientific feature type layer" identifies specific types of data such as grids, radial, and point data, and adds specialized methods for each of them. In detail, the "scientific feature type layer" adds functionality for data that is common in earth sciences and in particular provides a GeoGrid and a Grid Coordinate System. The GeoGrid is a specialized variable that explicitly handles X, Y, Z and time (T) dimensions while the Grid Coordinate System (on which the GeoGrid is based) provides a location in real physical space and time for the variables (usually with reference to the Earth). The Grid Coordinate System used by EST-WA at a minimum has a latitude (Lat) and longitude (Lon) coordinate axes, and optionally a time coordinate axis (T) and/or a vertical coordinate axis (Z).

The system is composed of two tools (see Fig. 8). The first one (EST-WA2D) allows to pre-view and pre-filter netCDF data through a 2D GUI (see Fig. 9 left). The second one (EST-WA3D) is instead the main module of the application. It is developed using the NASA World Wind Java SDK and allows visualization and analysis of data in a multidimensional scenario (see Fig. 9 right). A detailed description of EST-WA2D and EST-WA3D is provided in the following.

EST-WA2D allows the user to open any netCDF format file and read all of the data and metadata stored within it. Files can be loaded from local storage (e.g. the local file system of the computer where the application runs) or from remote Web servers accessed through the HTTP protocol. The HTTP access is viable thanks to a

---

[36]http://geomatica.como.polimi.it/elab/est-wa

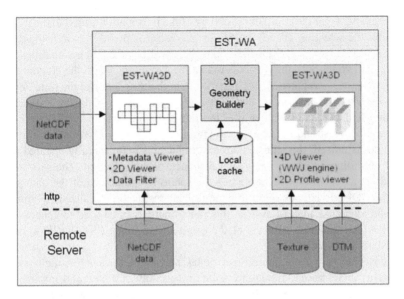

Fig. 8 EST-WA architecture

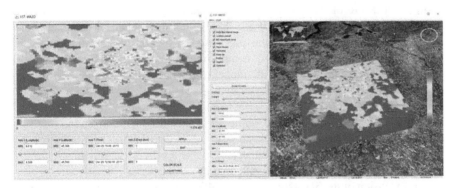

Fig. 9 EST-WA: the 2D viewer EST-WA2D (*left*) and the 3D viewer EST-WA3D (*right*)

feature of netCDF files which makes them accessible over the network by exploiting a simple Web server allowing for a range of requests for a resource. In this way the server must not return the data as "chunked" transfer-coding but allows a client to request only the data of interest.

For each variable stored in the netCDF file, the EST-WA2D GUI displays the data type, the measurement unit, the description, the rank, and the axes defining the variable domain. According to the variable type different information is shown. On each axis the measurement unit, the dimension, the step and the minimum/maximum values are reported. For each GeoGrid variable, the grid coordinate system information (i.e. the corresponding axes) is also given. Selecting each single element (variable, domain, grid or axis) it is possible to access the second level of metadata (attribute) and to get the complete overview of the information stored in the netCDF file. After the selection of the variable of interest, it is possible to activate the 2D map pre-viewer (see Fig. 9 left). Besides the usual browsing functionalities (pan, zoom-in and zoom-out), the dynamic map allows to interact with the content of the multidimensional structure by filtering data in both the spatial and the time domain.

For each axis (Lat, Lon, Z, and T) the GUI provides two sliders to select the minimum and maximum values (i.e. the range of the axis) in order to pre-filter the area of interest and select a spatio-temporal sub-domain. When the cursor of each slider is moved, the representation of the data is automatically updated in the viewer providing an immediate visual feedback of the selection. The sub-domain selection is a very useful feature when the spatio-temporal size of the data stored in the netCDF file is wide and the region of interest to be visualized is limited to a specific restricted area. Moreover, this functionality can be used to generate new netCDF files containing sub-sets of the original one, since EST-WA2D allows to export the filtering results.

EST-WA3D is a fully multidimensional netCDF file viewer which uses an innovative "doxel-based" modelling solution specially designed and implemented inside the platform. By extending the existing NASA World Wind Java classes to visualize the geometric primitives, new geometric figures were created. The basic 3D structure of the geometric model used by EST-WA is the voxel, which represents a cell of a regular grid in the three-dimensional space. The voxel is the 3D generalization of the rectangular pixel used in the 2D visualization and it is represented as a parallelepiped. The composition of all the geo-referenced voxels in a regular-grid frame forms the 3D model of the data stored in a netCDF file at a certain time. Each voxel is displayed with graduated colours (based on a colour ramp) depending on the value the variable takes at that position (see Fig. 9 right). When the volume occupied by the whole set of data does not correspond to a full parallelepiped (e.g. the 3D shape of the water filling an irregular natural basin) the voxels are not created in correspondence of the grid cells with "missing value" in order to obtain a full match of data with their representation.

The extension of the 3D static model to a 4D dynamic model, where time is also considered, was realized implementing a new geometric primitive called dynamic voxel, or simply doxel. A doxel is a voxel with the replacement of its static value with an array containing all the values the variable assumes at the different times. Compared to the static representation of a voxel (having a fixed colour), a doxel is displayed with different colours depending on the value the variable takes at that specific position at a certain time (dynamic representation). In order to also analyse the inner doxels of the multidimensional model, the volume can be sectioned with planes orthogonal to each other (see Fig. 10). For every axis (latitude, longitude and height) it is possible to define a minimum and a maximum cutting plane in order to visualize a slice of the original volume.

The viewer rendering is optimized by partitioning the 3D model into two logic sub-models: the inner part and the external layer, which is the shell of the model itself. Taking into account the selected cutting planes, the system sends to the

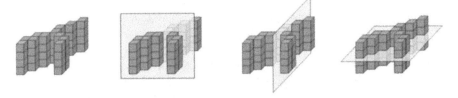

Fig. 10 Voxel structure and the cutting planes

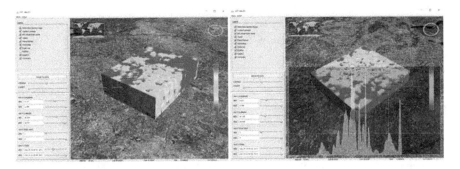

Fig. 11 EST-WA3D: sectioned model (*left*) and section profile graphic (*right*)

rendering engine only the filtered shell doxels (the shell doxels included in the domain delimited by the cutting planes) and the doxels which close the volume at locations corresponding to the cutting planes themselves. Compared to a basic solution where all the doxels are represented, the performance is greatly improved. Just to give an example, considering a cube of $100 \times 100 \times 100$ doxels, in the optimized case only 58,808 values are rendered instead of 1,000,000: the decrease of computation is equal to 94%. A local caching system finally increases the efficiency in accessing and handling the data using a separate management of the whole dataset (static dataset) and the current visible dataset (dynamic dataset) obtained from the insertion/removal of doxels depending on the cutting planes.

The environment is similar to the default NASA World Wind interface (on which it is based) but new tools for browsing data in time, analyzing the temporal evolution of the variable and slicing the data model to see the inner doxels have been implemented (see Fig. 11). Data can be sectioned along a constant longitude, latitude or height and, in order to increase the visual analysis of the Z (height) dimension, a vertical shift and a vertical exaggeration can be applied to the model. Moreover, a legend with the colour ramp corresponding to the variable values allows the quantitative analysis of the phenomenon (see Fig. 11 left). The value of a single voxel at a certain time and location or the values of all the voxels along a profile are readable thanks to an implemented 2D viewer appearing over the main map (see Fig. 11 right). Inside this semi-transparent window the graphic of the variable along the selected profile and the single values are dynamically synchronized with any movement of the scene and the consequent displacement of the profile along the 3D model. The user has full control of the profile, i.e. both translation and rotation in all directions are possible.

Finally, to completely customize the environment and to improve the spatio-temporal contextualization of the variable represented, functionalities for accessing WMS layers are also available. Through a dedicated GUI that can be activated from the menu bar, users can connect to any OGC compliant WMS server and select layers to be displayed on the virtual globe as background and/or auxiliary thematic maps.

Application of EST-WA to Explore Geo-Crowdsourced Data

Geo-crowdsourced data, which are constantly and often unconsciously contributed by users through wearable sensors, are to all effects comparable to the kinds of scientific variables mentioned above whose spatio-temporal analysis is suited for EST-WA exploitation. A specific application is hereafter presented which concerns citizen-sensed telecommunication data.

These data refer to mobile call data records in Milan, Italy, and were made available as open data by the Italian Telecom Italia mobile operator in the frame of the Telecom Italia Big Data Challenge.[37] Data have a temporal resolution of 10 min for a total of 2 months (November and December 2013) and a spatial resolution of 235 m × 235 m for a total of 100 × 100 cells covering Milan city. Five different phone activities are recorded: incoming and outcoming calls, incoming and outcoming SMSs (Short Message Services) and Internet connections. In the case of calls and SMSs, a Call Detail Record (CDR) is generated each time a user receives/issues a call or receives/sends an SMS. Regarding the Internet, a CDR is generated each time a user starts or ends a connection or, during the same connection, when there have been more than 15 min or 5 MB from the last generated CDR.

The available data, which include about 100M records, are stored in the NoSQL MongoDB[38] database from which, thanks to an ad hoc Python script, netCDF files can be generated on the fly.

These netCDF files are organized as four-dimensional regular grids, where the Z component is represented by the day (at most 61 slices corresponding to the 61 days) and the time component is represented by the 10 min interval (at most 144, corresponding to the count of 10 min timeslots in a day). The dimension of each doxel is therefore 235 m × 235 m × 1 d × 10 min.

The Python script to generate netCDF data is run through a Web application with a form that users have to fill.[39] Referring to time there are two main choices.

---

[37]http://theodi.fbk.eu/openbigdata

[38]https://www.mongodb.org

[39]http://landcover.como.polimi.it/BigNetCDF

Fig. 12 EST-WA visualization of SMSs sent in Milan city on December 31, 2013 at 12 pm (*left*) and 11 pm (*right*)

In the first case, users can select either all subsequent days or only Mondays, only Tuesdays, etc. from a starting date to an ending date; in the second case, users can select multiple dates from a calendar. Users have then to select the phone activity they are interested in (calls in, calls out, SMSs in, SMSs out, and Internet connections), submit the request and receive the file. Once the file has been sent to the user, it is deleted from the server filesystem to avoid its overload.

Generated netCDF files can be then spatio-temporally explored through the EST-WA interface. As an example, Fig. 12 shows the visualization of the SMSs out at two different hours of December 31, 2013: 12 pm (noon) and 11 pm. The colour ramp varies from blue (lowest counts of SMSs out) to red (highest counts). From the spatial perspective, the number of SMSs is in both cases higher in Milan city centre than in the suburbs. From the temporal point of view, the number of SMSs sent by people clearly shows a significant increase when midnight approaches, as this is typically the timeslot to send relatives and friends wishes for the new year. It is worth noting that the use of an intelligent multidimensional geo-viewer like EST-WA provides a much more meaningful and comprehensive understanding of the data compared to those achievable with traditional 2D or static 3D solutions.

Conclusions

This chapter has investigated the technological-scientific interaction between multi-dimensional visualization and the universe of VGI and citizen science. Although virtual globes applications and volunteered/contributed GI have both appeared almost a decade ago, their mutual interconnection is still a largely unexplored topic. Virtual globes have had the merit of democratizing GIS (Butler 2006); however 10 years later they remain strongly under-exploited for a number of scientific needs.

As shown in the literature review, VGI-related exploitation of virtual globes is often limited to the pure access and visualization of user-produced datasets (e.g. geotagged photos), with Google Earth—a primarily commercial product— still being the most popular alternative. User interaction is usually restricted to few, simple operations, the amount of accessible data is limited and the use of time dimension to navigate data is almost absent. Thus, validating what Goodchild et al. (2012) stated, still in 2016 we recognize how far the Digital Earth idea envisioned by Gore (1998) is.

In contrast to the current scenario, two advanced applications have been presented which prove the key role multidimensional applications can play when dealing with citizen-sensed data. The first application meets the need of exploiting virtual globes as general-purpose, VGI-based collaboration platforms. Besides retrieving and displaying user-collected data, the platform allows scientific communities to create and manage their own projects. The second application addresses the problem of spatio-temporally exploring complex datasets contributed by citizen sensors. Based on a scientific data format like netCDF, the platform acts as an intelligent geo-viewer able to easily spot macroscopic phenomena and eventually— at a more detailed level—even less evident patterns, which are more complex to study. In contrast to Google Earth, the use of an open source, scientifically oriented virtual globe like NASA World Wind has allowed to deeply customize and enrich the applications according to the needs.

The example of telecommunication data presented above shifts the discussion to the broad theme of big data, which recent literature has shown to be well-connected to VGI (see e.g. Fischer 2012). Despite statistical and processing techniques are required to perform full data analytics (e.g. to correlate telecommunication data to specific weekdays and/or weather/environmental conditions), a thorough visualization represents the first crucial step when dealing with big data.

Finally, it is worth mentioning the difficulties users sometime experience in exploiting complex multidimensional applications such as those described. While a true realization of DE—including robust handling of real-time data streams—would be desirable and beneficial for many scientific and societal needs, digital divide is still preventing a large portion of humankind from access and/or real exploitation. Therefore, there are at least two challenges to face: to reduce the complexity of such applications on one side and to educate people in using them on the other side. On both of them we are strongly working.

References

Annoni A, Craglia M, Ehlers M, Georgiadou Y, Giacomelli A, Konecny M, Ostlaender N, Remetey-Fülöpp G, Rhind D, Smits P, Schade S (2011) A European perspective on digital earth. Int J Dig Earth 4(4):271–284

Arsanjani JJ, Zipf A, Mooney P, Helbich M (eds) (2015) OpenStreetMap in GIScience – experiences, research and applications. Springer, Berlin

Bastin L, Buchanan G, Beresford A, Pekel JF, Dubois G (2013) Open-source mapping and services for Web-based land-cover validation. Eco Inform 14:9–16

Beltran A, Abargues C, Granell C, Núñez M, Díaz L, Huerta J (2013) A virtual globe tool for searching and visualizing geo-referenced media resources in social networks. Multim Tools Appl 64(1):171–195

Blenkinsop TG (2012) Visualizing structural geology: from Excel to Google Earth. Comput Geosci 45:52–56

Blower JD, Gemmell A, Haines K, Kirsch P, Cunningham N, Fleming A, Lowry R (2007) Sharing and visualizing environmental data using Virtual Globes. In Proceedings of the UK e-Science 2007 All Hands Meeting 2007, Nottingham, September 10–13 2007

Bodzin AM, Anastasio D, Kulo V (2014) Designing Google Earth activities for learning Earth and environmental science. In: Teaching science and investigating environmental issues with geospatial technology. Springer, Berlin, pp 213–232

Brovelli MA, Zamboni G (2012) Virtual globes for 4D environmental analysis. Appl Geomat 4:163–172

Brovelli MA, Zamboni G (2014) Environmental space and time web analyzer. In Proceedings of the 2014 Conference on Big Data from Space (BiDS'14), Research, Technology and Innovation (RT&I) (pp 224–226), Frascati, November 12–14, 2014

Brovelli MA, Minghini M, Zamboni G (2012) Valorisation of Como historical cadastral maps through modern web geoservices. In: ISPRS Annals of the Photogrammetry, Remote Sensing and Spatial Information Sciences, I-4, 287–292

Brovelli MA, Minghini M, Zamboni G (2014) Three dimensional volunteered geographic information: a prototype of a social virtual globe. Int J 3D Inform Model 3(2):19–34

Brovelli MA, Hogan P, Minghini M, Zamboni G (2013) The power of Virtual Globes for valorising cultural heritage and enabling sustainable tourism: NASA World Wind applications. In: International archives of the photogrammetry, Remote Sensing and Spatial Information Sciences, XL-4/W2, 115–120

Brovelli MA, Arias MC, Zamboni G (2015) From paper maps to the Digital Earth and the Internet of Places. Rendiconti Lincei 26(1):97–103

Butler D (2006) Virtual globes: the web-wide world. Nature 439(7078):776–778

Chien NQ, Tan SK (2011) Google Earth as a tool in 2-D hydrodynamic modeling. Comput Geosci 37(1):38–46

Clark ML, Aide TM (2011) Virtual interpretation of Earth Web-Interface Tool (VIEW-IT) for collecting land-use/land-cover reference data. Remote Sens (Basel) 3(3):601–620

Cohn JP (2008) Citizen science: can volunteers do real research? Bioscience 58(3):192–197

Cooper AK, Coetzee S, Kourie D (2010) Perceptions of virtual globes, volunteered geographical information and spatial data infrastructures. Geomatica 64(1):73–88

Craglia M, Goodchild MF, Annoni A, Camara G, Gould M, Kuhn W, Mark D, Masser I, Maguire D, Liang S, Parsons E (2008) Next-generation digital earth: a position paper from the Vespucci Initiative for the Advancement of Geographic Information Science. Int J Spat Data Infrastruct Res 3:146–167

Craglia M, de Bie K, Jackson D, Pesaresi M, Remetey-Fülöpp G, Wang C, Annoni A, Bian L, Campbell F, Ehlers M, van Genderen J, Goodchild M, Guo H, Lewis A, Simpson R, Skidmore A, Woodgate P (2012) Digital Earth 2020: towards the vision for the next decade. Int J Dig Earth 5(1):4–21

Elvidge CD, Tuttle BT (2008) How virtual globes are revolutionizing earth observation data access and integration. In: The International Archives of the Photogrammetry, Remote Sensing and Spatial Information Sciences, 37-B6a, 137–139

Fischer F (2012) VGI as Big Data: a new but delicate geographic data-source. GeoInform 15(3):46–47

Fritz S, McCallum I, Schill C, Perger C, Grillmayer R, Achard F, Kraxner F, Obersteiner M (2009) Geo-Wiki.Org: the use of crowdsourcing to improve global land cover. Remote Sens (Basel) 1(3):345–354

Fritz S, McCallum I, Schill C, Perger C, See L, Schepaschenko D, van der Velde M, Kraxner F, Obersteiner M (2012) Geo-Wiki: an online platform for improving global land cover. Environ Model Software 31:110–123

Gede M, Ungvári Z, Zentai L (2013) Virtual globes museum 2.0 – adding the power of community. In Proceedings of 26th International Cartographic Conference, Dresden, August 25–30, 2013

Goodchild MF (2007) Citizens as sensors: the world of volunteered geography. GeoJ 69(4):211–221

Goodchild MF (2008) The use cases of digital earth. Int J Dig Earth 1(1):31–42

Goodchild MF, Guo H, Annoni A, Bian L, de Bie K, Campbell F, Craglia M, Ehlers M, van Genderen J, Jackson D, Lewis AJ, Pesaresi M, Remetey-Fülöpp G, Simpson R, Skidmore A, Wang C, Woodgate P (2012) Next-generation digital earth. Proc Natl Acad Sci 109(28):11088–11094

Gore A (1998) The digital earth: understanding our planet in the 21st Century. Aust Surv 43(2):89–91

Guralnick RP, Hill AW, Lane M (2007) Towards a collaborative, global infrastructure for biodiversity assessment. Ecol Lett 10(8):663–672

Harvey F (2013) To volunteer or to contribute locational information? Towards truth in labeling for crowdsourced geographic information. In: Sui D, Elwood S, Goodchild M (eds) Crowdsourcing geographic knowledge. Springer, Berlin, pp 31–42

Harvey F (2009) More than names — digital earth and/or virtual globes. Int J Spat Data Infrastruct Res 4:111–116

Jones MT (2007) Google's geospatial organizing principle. IEEE Comput Graph Appl 27(4):8–13

Keysers JH (2015) Review of digital globes 2015. http://www.crcsi.com.au/assets/Resources/Globe-review-paper-March-2015.pdf. Accessed 3 Feb 2016

Kisilevich S, Mansmann F, Bak P, Keim D, Tchaikin A (2010) Where would you go on your next vacation? A framework for visual exploration of attractive places. In Proceedings of the Second International Conference on Advanced Geographic Information Systems, Applications, and Services (GEOprocessing 2010) (pp 21–26), St. Maarten, February 10–16, 2010

Lindner-Fally M, Zwartjes L (2012) Learning and teaching with digital earth – teacher training and education in Europe. GI Forum 2012:272–282

Loesch B, Christen M, Nebiker S (2012) OpenWebGlobe – an open source SDK for creating large-scale virtual globes on a WebGL basis. In: International Archives of the Photogrammetry, Remote Sensing and Spatial Information Sciences, XXXIX-B4, 195–200

Maguire DJ (2007) GeoWeb 2.0 and volunteered GI. In Workshop on Volunteered Geographic Information (pp 104–106), Santa Barbara, CA, December 13–14, 2007

Martínez-Graña AM, Goy JL, Cimarra CA (2013) A virtual tour of geological heritage: valourising geodiversity using Google Earth and QR code. Comput Geosci 61:83–93

Mooney P, Corcoran P (2014) Has OpenStreetMap a role in digital earth applications? Int J Dig Earth 7(7):534–553

Nebiker S, Bleisch S, Christen M (2010) Rich point clouds in virtual globes – a new paradigm in city modeling? Comput Environ Urban Syst 34(6):508–517

Nebiker S, Christen M, Eugster H, Flückiger K, Stierli C (2007) Integrating mobile geo sensors into collaborative virtual globes-design and implementation issues. In: Proceedings of the 5th International Symposium on Mobile Mapping Technology MMT'07, Padua, 29–31 May 2007

Norris P (2001) Digital divide: civic engagement, information poverty, and the internet worldwide. Cambridge University Press, Cambridge

O'Reilly T (2005) What is Web 2.0: design patterns and business models for the next generation of software. http://oreilly.com/pub/a/web2/archive/what-is-web-20.html. Accessed 1 Feb 2016

Peng ZR, Tsou MH (2003) Internet GIS. John Wiley & Sons, Hoboken, NJ

Plewe B (2007) Web cartography in the United States. Cartogr Geogr Inf Sci 34(2):133–136

Putz S (1994) Interactive information services using World-Wide Web hypertext. Comput Net ISDN Syst 27(2):273–280

Rew R, Davis G, Emmerson S, Davies H, Hartnett E, Heimbigner D (2010) The NetCDF users guide, data model, programming interfaces, and format for self-describing, portable data, NetCDF Version 4.1. Unidata Program Center

Sahr K, White D, Kimerling AJ (2003) Geodesic discrete global grid systems. Cartogr Geogr Inform Sci 30(2):121–134

Scharl A, Tochtermann K (eds) (2007) The geospatial web: how geobrowsers, social software and the web 2.0 are shaping the network society. Springer, London

Schroth O, Pond E, Campbell C, Cizek P, Bohus S, Sheppard SR (2011) Tool or toy? Virtual globes in landscape planning. Fut Int 3(4):204–227

Schultz RB, Kerski JJ, Patterson TC (2008) The use of virtual globes as a spatial teaching tool with suggestions for metadata standards. J Geogr 107(1):27–34

See L, Fritz S, Perger C, Schill C, McCallum I, Schepaschenko D, Duerauer M, Sturn T, Karner M, Kraxner F, Obersteiner M (2015) Harnessing the power of volunteers, the internet and Google Earth to collect and validate global spatial information using Geo-Wiki. Technol Forecast Social Change 98:324–335

Smith TM, Lakshmanan V (2011) Real-time, rapidly updating severe weather products for virtual globes. Comput Geosci 37(1):3–12

Stensgaard AS, Saarnak CF, Utzinger J, Vounatsou P, Simoonga C, Mushinge G, Rahbek C, Møhlenberg F, Kristensen TK (2009) Virtual globes and geospatial health: the potential of new tools in the management and control of vector-borne diseases. Geospat Health 3(2):127–141

Tomaszewski B (2011) Situation awareness and virtual globes: applications for disaster management. Comput Geosci 37(1):86–92

Valentini L, Brovelli MA, Zamboni G (2014) Multi-frame and multi-dimensional historical digital cities: the Como example. Int J Dig Earth 7(4):336–350

Vu HQ, Li G, Law R, Ye BH (2015) Exploring the travel behaviors of inbound tourists to Hong Kong using geotagged photos. Tour Manag 46:222–232

Webley PW (2011) Virtual Globe visualization of ash–aviation encounters, with the special case of the 1989 Redoubt–KLM incident. Comput Geosci 37(1):25–37

Wu H, He Z, Gong J (2010) A virtual globe-based 3D visualization and interactive framework for public participation in urban planning processes. Comput Environ Urban Syst 34(4):291–298

Yovcheva Z, van Elzakker CP, Köbben B (2010). *Collaborative mapping and spatio-temporal data dissemination through a web-based virtual globe application.* In: Proceedings of 3rd International Conference on Cartography and GIS, Nessebar, 15–20 June 2010

Yu L, Gong P (2012) Google earth as a virtual globe tool for Earth science applications at the global scale: progress and perspectives. Int J Remote Sens 33(12):3966–3986

Zhu LF, Wang XF, Zhang B (2014) Modeling and visualizing borehole information on virtual globes using KML. Comput Geosci 62:62–70

7

The Changing Landscape of Geospatial Information Markets

Conor O'Sullivan, Nicholas Wise, and Pierre-Philippe Mathieu

Abstract We live in an increasingly global, connected and digital world. In less than a decade or so, fast developments in digital technologies, such as the Cloud, Internet, wireless network, and most importantly mobile telephony, have dramatically changed the way we work, live and play. Rapid advances in Information and Communication Technologies (ICT) foster a new world of cross-disciplinary data-intensive research characterised by openness, transparency, access to large volume of complex data, availability of community open tools, unprecedented level of computing power, and new collaboration among researchers and new actors such as citizen scientists. Identifying and understanding the key drivers of change in the data economy and EO sector (including technological, human, cultural and legal factors) is essential to providing context on which to build an EO strategy for the twenty-first century. The emergence of cloud computing is already transforming the way we access and exploit data. This has led to a paradigm shift in the way to distribute and process data, and in creating platforms that drive innovation and growth in user applications.

Introduction

The Earth Observation industry, part of the wider data economy, is experiencing a number of factors that are driving change across the value chain. These include, to name a few, leveraging IT infrastructure such as cloud computing, the rise of platforms and the Internet of Things (IoT), interconnected terrestrial and space-borne systems, diversification of business models and open data policies. Copernicus, the European flagship programme to provide geo-information services to EU policy makers, provides a strong opportunity as market driver for EO-based services. According to a recent survey by the European Association of Remote

C. O'Sullivan (✉) • N. Wise
Satellite Applications Catapult, Harwell, UK
e-mail: Conor.OSullivan@sa.catapult.org.uk; nicholas.wise@sa.catapult.org.uk

P.-P. Mathieu
ESA/ESRIN, Frascati, Italy
e-mail: Pierre.philippe.mathieu@esa.int

Sensing Companies (EARSC), Industry is optimistic about the positive impact the Copernicus programme will have on their business (EARSC 2015).

The European Commission's Digital Single Market Package is a genuine driver for EU growth and new jobs. It highlights the benefits of a stronger Digital Single Market and its potential for higher growth and new jobs, and increasing global competitiveness:

> *Full and efficient exploitation of tools and services such as Cloud Computing, Big Data, Automation, Internet of Things and Open Data can drive for better productivity and better services, and therefore should be facilitated, including through market driven solutions, R&D and the promotion of the necessary skills and capacity building, along with further ICT standardisation and interoperability* (Council of the European Union 2015)

The volume, variety and velocity of data are increasingly rapidly and "Big Data" acts as the oil in the supply chain for many industries. Within the next few years, ESA spacecraft alone will obtain approximately 25 PB of Earth Observation (EO) data as a result of the Copernicus programme (Di Meglio et al. 2014). In addition, data is generated from a multitude of sources, including small satellite constellations, ground and airborne sensors (e.g. Unmanned Aerial Vehicles, UAVs), social media, machine to machine (M2M) communications and crowdsourcing.

The cost-effective to process and store data is falling, making it simpler and more economical to capture larger datasets by leveraging the significant investment made by companies in the cloud computing industry. Increasing value lies in turning this data into knowledge and actionable insights, thereby enabling new applications and services that create value for end users. With views into daily activity being refreshed at a faster rate than ever before, just selling raw pixels is not enough to satisfy end-user demands, those pixels need to be turned into insights. This is evident in the EO sector where ambitious start-ups, such as Planet, are building constellations of small satellites and developing cutting-edge analytics to extract value from the data captured. Many of these start-ups consider themselves as satellite powered data companies. In Planet most recent round of funding the company plans to use the investment to develop its capabilities for processing, interpreting and selling data contained in its images. It was this focus that attracted interest from Data Collective, a venture capital firm, which has backed several big data start-ups (Financial Times 2015).

Data value chain. Source: Digital Catapult (2014)

More EO missions are being launched than ever before. Reduced launch costs, miniaturisation of technology, improved on-board processing and better reliability are driving increased interest in small satellites by new commercial companies. To unlock the economic potential of data from the increasing number of satellites, public agencies and private companies are creating data products that aim to be responsive to user needs. The satellite data generated from ESA's Sentinel satellite constellation, for example, will provide actionable insights from the observation of the planet thanks to an array of sensor technologies, including Synthetic Aperture Radar (SAR) and Multispectral/Hyperspectral sensors.

EO as a Platform turns raw data into knowledge through processing and analysis, creating value within and across various sectors. EO and remote sensing data has significant potential to help us manage the modern world and our planet's resources. Applications and services are already emerging for emergency response and environmental monitoring, while emerging markets such as precision agriculture, monitoring of illegal fishing and management of natural resources are rapidly developing. There is increasing value to be created by reaching more customers through the applications of big data. The EO data value chain creates opportunity for small and medium sized enterprises (SMEs) and start-ups to engage with the space sector, and generate value from satellite missions by developing applications for citizens, local government and commercial industry.

Public agencies are increasingly interested in how they can interact effectively with companies that have enabled a globally distributed applications ecosystem and are investing extensively in cloud computing infrastructure. Commercial cloud providers, like Microsoft Azure and Amazon Web Services, are key enablers of building, deploying and managing scalable applications with the aim of reaching a global audience. Open data policies can enable the private sector to do just that, and reach a wide audience of application developers and end users. According to The Economist, information held by governments in Europe could be used to generate an estimated €140 billion worth of value a year (The Economist 2013). In short, making official data public will spur innovation and create new applications. These benefits need to be balanced against the rights of individuals to have their data protected by government.

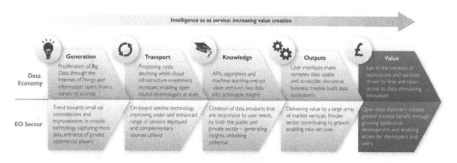

Summary: value chain, key drivers of change. Source: Satellite Applications Catapult

The key drivers of change in the data economy impacting the EO market include:

- *Rise of the platforms*: leveraging cloud computing infrastructure to process more and more layers of data, from multiple sources. Simplifying applications development and building an app ecosystem around scalable, on-demand IT infrastructure.
- *Data as a Service*: user manages the application, everything else is delivered as a service. Moving users closer to the data ("data gravity") via Content Delivery Networks (CDNs).[1]
- *Open data policies*: demand from users and government policies changing towards improved access to data and tools.
- *New business models*: growing an ecosystem of researchers and developers so that people can easily gain access to and use a multitude of data analysis services quickly, through cloud and high performance computing (HPC) platforms, to add knowledge and open source tools for others' benefit.
- *Sensor use growing*: Internet of Things and sensors intelligently working at the edge of networks, complementarity of space-borne and terrestrial data.
- *Crowdsourcing*: citizen science platforms and their commercial capability.
- *Disruptive innovation*: introduces a new value proposition. They either create new markets or reshape existing ones.

Rise of the Platforms

Cloud computing refers to accessing highly scalable computing resources through the Internet, often at lower prices than those required to install on one's own computer because the resources are shared across many users. Cloud computing has become the next logical evolution in computing—combining the critical elements of each architecture that came before it.

The NIST (National Institute of Standards and Technology) offers the following definition of cloud computing:

> *Cloud computing is a model for enabling ubiquitous, convenient, on-demand network access to a shared pool of configurable computing resources (e.g. networks, servers, storage, applications, and services) that can be rapidly provisioned and released with minimal management effort or service provider interaction.*

Cloud computing is about the capability to access any information, at any time, regardless of the resources required and the location of the infrastructure, data,

---

[1]CDNs: a content delivery network or content distribution network (CDN) is a large distributed system of servers deployed in multiple data centres across the Internet. The goal of a CDN is to serve content to end-users with high availability and high performance. Examples include Microsoft Azure and Amazon CloudFront.

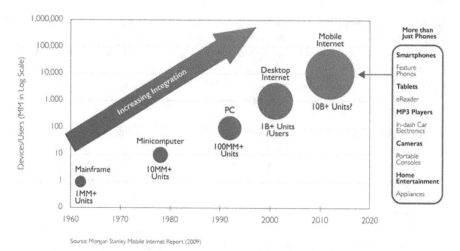

Each new computing cycle typically generates around 10x the installed base of the previous cycle

Each new computing cycle typically generates around 10× the installed base of the previous cycle. Source: Kleiner Perkins Caufield Buyers (2014) *Internet Trends 2014*, see www.kpcb.com/InternetTrends retrieved on December 5th 2014

application or user. The availability of robust cloud platforms and applications have begun to enable businesses to shift budget spending from capital expense (including dedicated, physical on-site servers) to operating expense (shared hosting by cloud providers) (Woodside Capital Partners 2014).

Cloud computing services are typically segmented into three different areas:

1. *Infrastructure as a Service (IaaS)*—third-party provider hosts virtualised computing resources over the Internet, through which customers can host and develop services and applications.
2. *Platform as a Service (PaaS)*—used by developers to build applications for web and mobile using tools provided by the PaaS provider—these range from programming languages to databases.
3. *Software as a Service (SaaS)*—software is hosted in the cloud, but appears on devices with full functionality.

Access to the Cloud and platforms are required to capitalise on the data and tools being created to make it easier and faster to discover, process and action on EO datasets. The cloud provides scalability and flexibility in a cost-efficient manner. There should be a combination of ESA and cloud providers supporting the communities within the EO ecosystem to develop tools and ensure data access.

Commercial cloud, capitalising on co-location of computing resources and data storage, is now becoming widely adopted as it offers several advantages enabling users to (1) perform data-intensive science, (2) ensure traceability of workflows,

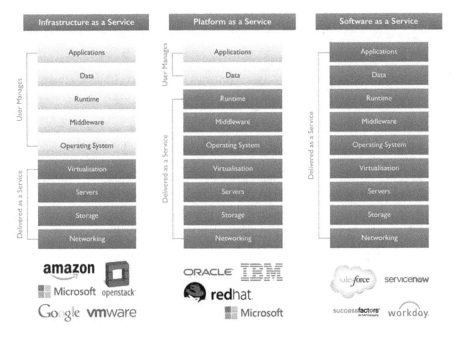

Cloud delivery options. Source: Woodside Capital Partners (2014) *OpenStack: Is this the Future of Cloud Computing?* see www.woodsidecap.com/wp-content/uploads/2014/11/WCP-OpenStack-Report_FINALa.pdf. Retrieved on 21st November 2014.

input data sets and therefore enabling to reproduce results, (3) facilitate integration with other non-EO data through standard web services (e.g. open data and IoT for smart cities) and (4) open new business models, whereby commercial data, proprietary software, apps and computing resources can now be rented on-demand (as opposed to purchased) to generate Information as a Service. Given the number of EO platforms that are being developed and coming on-line, the challenge revolves around the technical and economic aspects of interoperability, such as:

- A common EO data pool from all EO missions in Europe;
- A processing capability management (sharing of resources and cloud services for processing);
- Sustainability through fair and democratic access to all resources (data, intellectual properties, enabling technologies/computing) by means of underlying implementation principles based on brokerage within an open environment;
- Federated user interfaces subsystems (e.g. interlinked EO data catalogues), interface and standards definition and agreement;
- Development of common value-creation techniques (research in data analytics and information retrieval, information visualization, data mining, fusion of in situ data with geo-information, etc.);

- Definition and application of commonly agreed data management principles to ensure data discoverability, accessibility, usability, preservation and curation.
- With the emergence of cloud-based storage and computing, users are now able to easily and cheaply process data on demand. This leads to a paradigm shift whereby data providers are now moving the computations—and therefore the users—to the data, rather than the data to the users. In this context, ESA is developing a series of EO Thematic application platforms or Virtual Research Environments (e.g. Geohazard SuperSites) to exploit the data for different applications and communities.
- Increasing data access and data liquidity, through interaction with the wider data economy and embracing ICT trends (Cloud, open software, APIs), will support new products with market opportunity. This will pave the way for new and existing organisations that have valuable skills or market knowledge, to enter the EO community by making it easier to discover, access and process EO datasets. Both the "democratisation" of GIS skills and making EO tools and interfaces more user friendly are crucial to enabling a much wider range of users.

Core to the data access agenda is cloud computing, which allows users to use on-demand computing resources co-located with data staged for analysis in the cloud. Commercial cloud providers, such as Amazon Web Services and Microsoft Azure, offer various tools and techniques to enable the build, deployment and management of scalable applications, with the aim of reaching a global audience. Through these platforms, underlying computer and storage resources scale automatically to match application demand. When data is made publicly available via a commercial cloud provider, users can access and analyse the data extremely rapidly and at very large scale.

Core also to increasing data access and liquidity are powerful functional interfaces (APIs) that allow discovery and retrieval of EO data and products, thereby delivering analysis ready data. The cloud based API Economy is key to accelerating value, improving performance, and extending products and services to the widest possible audience. User friendly API's are essential—they must be flexible enabling enhanced features so entrepreneurs and developers can use combinations of data for specific themes. A well-established way to simplify access to services and data is to implement an API since this enables a developer to exploit their specialist knowledge to create higher level products and services without having to invest a large amount of time and effort to access the relevant data and create, and then verify, basic services. Companies like Planet Spire are following this approach by developing APIs and allowing "plug-and-play" access to them. The biggest player in the satellite data industry, DigitalGlobe has recently announced the launch of Maps API to allow software developers to embed satellite images, maps and other geospatial content into their mobile and web applications.

Data as a Service

More recently, the concept of data as a service (DaaS) has developed. It represents the enablement of regular, non-expert users to effectively take control of often highly complex and traditionally inaccessible IT tools. DaaS can be defined as the sourcing, management, and provision of data delivered in an immediately consumable format to users.[2] Like all members of the "as a Service" family, DaaS is based on the concept that the product, data in this case, can be provided on demand to the user regardless of geographic or organisational separation of provider and consumer. Data quality can happen in a centralised place, cleansing and enriching data and offering it to different systems, applications or users, irrespective of where they were in the organization or on the network.

An increasing number of Internet network owners have built their own content delivery networks (CDNs) to improve on-net content delivery and to generate revenues from content customers. For example Microsoft builds its own CDN in tandem with its own products through its Azure CDN. CDNs, like Azure and Amazon's CloudFront, are key enablers of building, deploying and managing scalable applications with the goal of reaching a global end user audience. Through these platforms, underlying computer and storage resources scale automatically to match application demand.

According to some reports, 300,000 APIs (Application Programming Interfaces) are projected to be registered by 2020.[3] APIs are the fastest growing, business-influencing technology in the IT industry today. With an API, developers can exploit functions of existing computer programmes in other applications. Companies are exposing APIs to allow others to consume their business functions, for a profit. Where Windows and Linux have been traditional development platforms of the past, Google, Facebook, Twitter and other companies are becoming the development platforms of the future. All of these companies built a functional platform of business capabilities and extended their business models by exposing APIs so that developers can exploit their functionality. Google Maps is a key example. Many developers write mash-ups (using content for an application from more than one source) on top of Google Maps for various reasons, for example retail store locators, traffic reports, road conditions and so on.

APIs are now coming of age with the advent of cloud computing, where the ability to host external APIs has matured to a point where cloud service providers have scalable capacity to handle transaction loads and spikes in traffic. Mobile platforms now put application reach on millions of devices, all having access to back-end APIs across the Internet. Amazon Web Services (AWS) Marketplace (Amazon's API marketplace) attracts not only developers and partners looking

---

[2] Oracle (2014) *Data-as-a-service: the Next Step in the As-a-service Journey,* see www.oracle.com/us/solutions/cloud/analyst-report-ovum-daas-2245256.pdf retrieved on 2nd March 2015.

[3] IBM (2013) *Global Technology Outlook.*

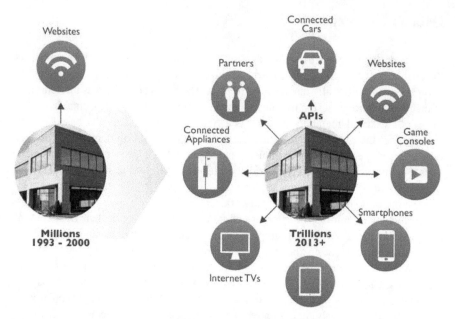

APIs are enabling more and more devices to connect. Source: IBM (2014) *Exposing and Managing Enterprise Services with IBM API Management.*

to exploit Amazon's APIs, but other vendors also, such as SAP and Oracle (that provide their own APIs on AWS, to offer analytics for additional value).

Open Data Policies

Data has been referred to as the new raw material of the twenty-first century. Like many other raw materials, it needs investment to locate, extract and refine it before it yields value. Open data, employed in combination with open platforms, such as APIs, expands the network of minds and unlocks the data's latent potential. As a result of increased demand for access to free data, governments and agencies are doing more to open up large amounts of public sector information to the public. ESA, for example, is implementing a free, full and open data policy through the Copernicus programme of Sentinel satellites.

In 1983, President Ronald Reagan made America's military satellite-navigation system, GPS, available to the world; entrepreneurs pounced on this opportunity. Car navigation, precision farming and three million American jobs now depend on GPS.[4] Official weather data are also public and avidly used by everyone from insurers to ice-cream sellers. All data created or collected by America's federal government must now be made available free to the public, unless this would violate privacy, confidentiality or security.

Open and machine-readable is the new default for government information. (US President, Barack Obama (2013))[5]

Many countries have moved in the same direction. In Europe, the information held by governments could be used to generate an estimated €140 billion a year.[6] McKinsey estimates the potential annual value to Europe's public sector administration at €250 billion.[7]

The emerging open data 'marketplace'

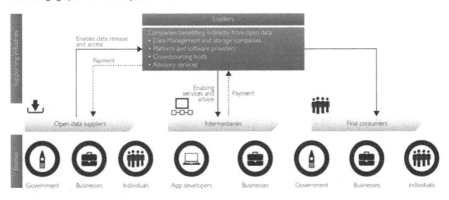

The emerging open data 'Marketplace'. Source: Deloitte (2012) *Open growth: Stimulating demand for open data in the UK,* see www2.deloitte.com/uk/en/pages/deloitte-analytics/articles/stimulating-demand-for-open-data-in-the-uk.html retrieved on 8th February 2015.

There are lots of companies, charities and individuals who would benefit if all the data the public sector holds was shared with them, particularly if it was shared only with them. However, those benefits have to be balanced against the rights of individuals to have their data protected by government, and the risks to individuals and to society of too much data being available (for example, through making fraud easier).

---

[4]The Economist (2013) *A new goldmine; Open data,* see www.economist.com/news/business/21578084-making-official-data-public-could-spur-lots-innovation-new-goldmine retrieved on 8th February 2015.

[5]Ibid.

[6]Ibid.

[7]McKinsey Global Institute (2011) *Big data: The next frontier for innovation, competition, and productivity.*

In *The Cathedral & the Bazaar*, a book by Eric on software engineering methods, the image of a bazaar is used to contrast the collaborative development model of open source software with traditional software development. In the traditional software development "vending machine model", the full menu of available services is determined beforehand. A small number of vendors have the ability to get their products into the machine, and as a result, the choices are limited, and the prices are high. A bazaar, by contrast, is a place where the community itself exchanges goods and services.[8]

In the technology world, the equivalent of a thriving bazaar is a successful platform. In the computer industry, the innovations that define each era are frameworks that enabled a whole ecosystem of participation from companies large and small. The personal computer was such a platform and so was the World Wide Web. This same platform dynamic is playing out now in the recent success of the Apple iPhone. Where other phones have had a limited menu of applications developed by the phone vendor and a few carefully chosen partners, Apple built a framework that allowed virtually anyone to build applications for the phone, leading to an explosion of creativity, with more than 100,000 applications appearing in little more than 18 months, and more than 3000 new ones now appearing every week.[9] Android, with a global smartphone operating system market share of around 80%,[10] is open-source software for a wide range of mobile devices and a corresponding open-source project led by Google. These successes are due to the openness around frameworks.

As applications move closer to the mass of data, for example by building an applications ecosystem around free and public data sets, more data is created. This concept of 'data gravity' is about reducing the cycle time/feedback loop between information and the data presented. This is achieved through lower latency and increased. There is an accelerative effect as applications move closer to data.[11]

Smartphones and tablets, collectively "mobile devices", are the fastest adopted technology in history. They have been adopted faster than cell phones, personal computers (PCs), televisions, even the Internet and electricity. The reason why the likes of Apple (iOS) and Google (Android) lead the way in mobile applications is because they combine a large pool of talented mobile developers with a robust development infrastructure. Apple ignited the app revolution with the launch of the App Store in 2008, and since then, an entire industry has been built around app design and development. According to recent announcements from Apple, apps on iOS generated over €8 billion in revenue for developers in 2014 and to date, App

---

[8]Lathrup, D. and Ruma, L. (2010) *Open Government: Collaboration, Transparency, and Participation in Practice*, O'Reilly Media.

[9]Ibid.

[10]Business Insider website, see www.businessinsider.com/iphone-v-android-market-share-2014-5?IR=T retrieved on 3rd March 2015.

[11]Dave McCrory (2011), Gathering Moss, Data Gravity, and Context, see www.datagravity.org retrieved on 17th March 2015.

Store developers have earned a cumulative €20 billion from the sale of apps and games.[12] According to Flurry Analytics, a mobile analytics firm, in 2014 overall app usage grew by 76%.

New Business Models

New entrants to the EO sector, including Planet and Spire, are opening their data to developers and end-users through APIs. APIs can make it easier to access EO data and to extract the embedded value. Planet has announced that it will release a developer API this year.[13]

Business models are also emerging to develop a more integrated network of stakeholders. CloudEO, a German company that supplies EO data on a pay-per-use or subscription basis,[14] aims to bring together imagery providers, analytics companies and customers through one platform. In order to attract and expand the user community beyond the boundaries of EO, the development of semantic search structures can play a pivotal role in reaching new users. The GEOinformation for Sustainable Development Spatial Data Infrastructure (GEOSUD SDI) is one example of this.[15]

Among the new products and services that are being developed, EO video data products are worth highlighting. Enabled by more frequent revisit times of EO satellite constellations, these products have the potential to improve the value proposition of a satellite data provider in applications such as disaster relief, surveillance and other applications that could benefit from real-time monitoring.[16] Canadian company, UrtheCast, has been granted the exclusive right to operate two cameras on the Russian module of the International Space Station (ISS).[17] As the ISS passes over the Earth, UrtheCast's twin cameras capture and download large amounts of HD (5 m resolution) video and photos. This data is then stored

---

[12]Apple (2015) see www.apple.com/pr/library/2015/01/08App-Store-Rings-in-2015-with-New-Records.html retrieved on 19th January 2015.

[13]Planet Labs website, see www.planet.com/flock1/ retrieved on 29th January 2015.

[14]Henry, C. (2014) *CloudEO Starts 'Virtual Constellation' Access with Beta Online Marketplace*, see www.satellitetoday.com/technology/2014/03/26/cloudeo-starts-virtual-constellation-access-with-beta-online-marketplace/ retrieved on 28th January 2015.

[15]M. Kazmierski *et al* (2014) *GEOSUD SDI: accessing Earth Observation data collections with semantic-based services*, see www.agile-online.org/Conference_Paper/cds/agile_2014/agile2014_138.pdf retrieved on 19th January 2015.

[16]Northern Sky Research (2013) *Satellite-Based Earth Observation, 5th Edition.*

[17]UrtheCast (2013), see www.investors.urthecast.com/interactive/lookandfeel/4388192/UrtheCast-Investor-Deck.pdf retrieved on 29th January 2015.

and made available via APIs on the basis of a pay-for-use model.[18] One of the innovative characteristics of UrtheCast's business model is the way it approaches the revenue streams it can tap into, for example by providing videos free of charge and generating an online advertising-like revenue from companies that will have their logos featured on the video in relation to their locations.[19]

Sensor Use Growing

The IoT connects sensors on items, products and machines, enabling users to receive a more fine-grained picture of information systems. IoT represents the next evolution of the Internet, taking a huge leap in its ability to gather, analyse, and distribute data that can be turned into information, knowledge, and actionable insights.[20]

The IoT is forecast to reach 26 billion installed units by 2020, up from 900 million 5 years ago.[21] Whether used individually or, as is increasingly the case, in tandem with multiple devices, sensors are changing our world for the better— be it by reminding us to take our medicine, or by tracking traffic flow. Satellite imaging of weather systems, vegetation changes, and land and sea temperatures can be combined with temperature and pollution data on the ground to provide a picture of climate change and man's impact on the planet. Limited range local sensors can provide detailed information that can be cross referenced with satellite data to validate models, which in turn can be used to provide wide area predictions and forecasts. This has been fundamental to the development of weather forecasting, and will be equally fundamental to many other satellite applications.

Embedding sensors in physical objects like computers, watches and robots, provides data to develop technologies that solve our needs and make business cases. For example, an imminent increase in the number of intelligent devices available is set to make supply chains smarter than ever. However it is not just information about the location of physical assets that will boost supply chain visibility. Data about their condition and state will be important, too. For example, if the temperature that food

[18]IAC (2014) *UrtheCast is #DisruptiveTech, Onwards and Upwards Blog*, see www.blog.nicholaskellett.com/2014/10/03/iac-2014-urthecast-is-disruptivetech/ retrieved on 19th January 2015.

[19]UstreamTV (2012) *UrtheCast Business Model*, see www.ustream.tv/recorded/26973814 retrieved on 19th January 2015.

[20]Cisco (2011) *The Internet of Things: How the Next Evolution of the Internet Is Changing Everything.*

[21]Financial Times (2014) *The Connected Business*, see www.im.ft-static.com/content/images/705127d0-58b7-11e4-942f-00144feab7de.pdf retrieved on 21st November 2014.

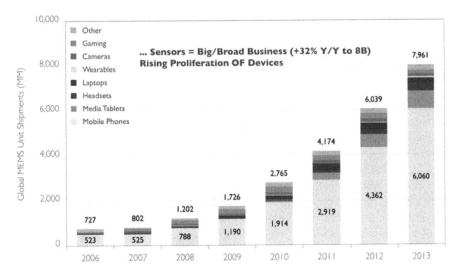

Rising proliferation of devices. Source: Kleiner Perkins Caufield Buyers (2014), *Internet Trends 2014*, see www.kpcb.com/InternetTrends retrieved on December 5th 2014.

products are kept at throughout the supply chain can be tracked, food companies have a better chance of extending shelf-life and reducing waste.[22]

Toyota has announced the development, in Japan, of the "Big Data Traffic Information Service", a new kind of traffic-information service utilising big data including vehicle locations and speeds, road conditions, and other parameters collected and stored via telematics services. Based on such data, traffic information, statistics and other related information can be provided to local governments and businesses to aid traffic flow improvement, provide map information services, and assist disaster relief measures.[23]

Crowdsourcing

There has been an explosion of activity in the area termed citizen science, crowdsourcing and volunteered geographic information (VGI). EO data is contributing to problem solving on a global scale. Some of the highest profile successes happen when this data is used in citizen science projects, where the power of large numbers of humans getting involved can achieve results that are simply not possible with computers alone:

---

[22] Financial Times (2014) *The Connected Business*, see www.im.ft-static.com/content/images/705127d0-58b7-11e4-942f-00144feab7de.pdf retrieved on 21st November 2014.

[23] ABI/INFORM Global (2013) *JAPAN: Toyota to launch Big Data Traffic Information Service*.

- Speedy provision of information to rescue and support services following natural or manmade disasters;
- Identifying evidence of illegal activities, including poaching and people trafficking;
- Improving maps and monitoring the environment.

Citizen science and crowdsourcing are being used in a very diverse range of applications, including archaeology, medicine, mapping, astronomy and disaster response. Science and mapping-related projects may either be organised through one of the aggregator platforms, such as Tomnod and Zooniverse, or stem from calls from organisations such as the UN and scientific bodies.

The Geo-Wiki project, for example, engages people in visual validation of maps, with a focus on land cover and land use. The tool can bring together datasets to be viewed on top of high resolution satellite imagery and has already been used to challenge assumptions about land available for biofuels. The tool also serves as a visualisation platform, bringing together global land cover datasets in a single place, which can be viewed on top of very high resolution satellite imagery from Google Earth and Bing.

The growth of the citizen science movement represents a major shift in the social dynamics of science, in blurring the professional/amateur divide and changing the nature of the public engagement with science.[24] There has been a shift from observing science passively to being actively engaged in the scientific discussion.

Citizen science encompasses many different ways in which citizens are involved in science. This may include mass participation schemes in which citizens use smartphone applications to submit wildlife monitoring data, for example, as well as smaller-scale activities like grassroots groups taking part in local policy debates on environmental concerns over fracking. Companies are forming to promote citizen science projects. Zooniverse, a citizen science web portal owned and operated by the Citizen Science Alliance, is home to the Internet's largest collection of citizen science projects. Zooniverse has well over one million volunteers that can get involved in projects to participate in crowdsourced scientific research. Unlike many early Internet-based citizen science projects (such as SETI@home) that used spare computer processing power to analyse data, known as volunteer computing, Zooniverse projects require the active participation of human volunteers to complete research tasks. By combining machine computing with human computing (digital volunteers), a more comprehensive analysis can be performed.

Projects have been drawn from disciplines including astronomy (the Galaxy Zoo project), ecology, cell biology, humanities, and climate science. Galaxy Zoo enables users to participate in the analysis of imagery of hundreds of thousands of galaxies drawn from NASA's Hubble Space Telescope archive and the Sloane Digital SkySurvey. It was started in 2007 by an Oxford University doctoral student, who decided to involve the community of amateur astronomers by using crowdsourcing.

---

[24]The Royal Society (2012) *Science as an Open Enterprise,* The Royal Society Science Policy Centre report.

To understand how these galaxies formed, astronomers classify them according to their shapes. Humans are better at classifying shapes than even the most advanced computer. More than 320,000 people have now taken part in Galaxy Zoo, over 120 million classifications have been completed, and there are now more than 25 peer-reviewed publications based on data from Galaxy Zoo.[25]

Disruptive Innovation

Defined by Clayton Christensen, disruptive innovation points to situations in which new organisations can use relatively simple, convenient, low cost innovations to create growth and successfully compete with incumbents. The theory holds that incumbent companies have a high probability of beating entrant firms when the contest is about *sustaining* innovations. However, established companies almost always lose to entrants armed with *disruptive* innovations.[26]

Sustaining innovations are what move companies along established improvement trajectories. They are improvements to existing products or dimensions historically valued by customers. Airplanes that fly farther, computers that process faster and cellular phone batteries that last longer are all examples of sustaining innovations.

Disruptive innovations introduce a new value proposition. They either create new markets or reshape existing ones. Christensen defines two types of disruptive innovation:

1. *Low-end* disruptive innovations can occur when existing products and services are "too good" and hence overpriced relative to the value existing customers can use. Nucor's steel mini-mill, Wal-Mart's discount retail store and Dell's direct-to-customer business model were all low-end disruptive innovations.
2. *New-market* disruptive innovations can occur when characteristics of existing products limit the number of potential consumers or force consumption to take place in inconvenient, centralised settings. The Sony transistor radio, Bell telephone, Apple PC and eBay online marketplace were all new-market disruptive innovations. They all created new growth by making it easier for people to do something that historically required deep expertise or significant wealth.

---

[25]Wikipedia, see www.en.wikipedia.org/wiki/Zooniverse_%28citizen_science_project%29 retrieved on 18th January 2015.

[26]Christensen et al (2004) *Seeing what's next: using the theories of innovation to predict industry change,* Harvard Business School Press.

Cloud Computing

Cloud-delivered enterprise solutions fit Christensen's concept of disruptive innovation. They offer cheaper, simpler and often more broadly applicable alternatives to legacy models of enterprise computing. They tend to start out as low-end disruptors, bringing cost and performance advantages to over-served customers, but as these technologies mature in their reliability and sophistication, they're spreading throughout organisations and solving some of the most demanding problems.[27]

In the 1990s, the enterprise software industry went through an upheaval as the client-server model displaced the mainframe. This new (and now old) standard represented large shift in value, because applications could now be much more powerful and modern using PC standards, data could mostly be centralised, and everything would run at a fraction of the cost compared to mainframes. Fast forward a decade and a half, and the same large-scale change has occurred yet again with most core applications being brought back to the web.

Distributing technology over the web offers current market leaders no intrinsic advantage that a start-up cannot access—that is, the web is served up democratically, whereas software in the past was usually delivered via partners or vendors with the most extensive salesforces. Cloud solutions generally embrace a world defined by collaboration, mobility, and openness. Many cloud solutions today are similarly disrupting incumbents by initially slipping into the "just good enough" category. Product roadmaps then become more comprehensive and customers are served in more meaningful ways.

Business Models in Cloud Computing

The following organisations are key players in the cloud computing market and in enabling a globally distributed applications infrastructure.

Microsoft Azure

Microsoft Azure delivers general purpose platform as a service (PaaS), which frees up developers to focus only on their applications and not the underlying infrastructure required. Having the IT infrastructure, hardware, operating systems and tools needed to support an application opens up possibilities for developers. The Microsoft hybrid cloud leverages both on-premises resources and the public cloud. Forty percent of Azure's revenue comes from start-ups and independent software vendors (ISVs), and 50% of Fortune 500 companies use Windows Azure. Microsoft

---

[27]Fortune (2011) see www.fortune.com/2011/09/27/is-the-cloud-the-ultimate-disruptive-innovation/ retrieved on 20th January 2015.

has invested $15 billion to build its cloud infrastructure, comprised of a large global portfolio of more than 100 datacentres, one million servers, content distribution networks, edge computing nodes, and fibre-optic networks.[28]

Leveraging Microsoft's significant investment in infrastructure and the Azure platform, NASA was able to more easily build and operate its new "Be a Martian" site—an educational game that invites visitors to help the space agency review thousands of images of Mars. Site visitors can pan, zoom and explore the planet through images from Mars landers, roving explorers and orbiting satellites dating from the 1960s to the present. In keeping with the rise of gamification, the site is also designed as a game with a twofold purpose: NASA and Microsoft hope it will spur interest in science and technology among students in the US and around the world. It is also a crowdsourcing tool designed to have site visitors help the space agency process large volumes of Mars images. Researchers at the NASA Jet Propulsion Laboratory (NASA/JPL) wanted to solve two different challenges—providing public access to vast amounts of Mars-related exploration images, and engaging the public in activities related to NASA's Mars Exploration Programme. The sheer volume of information sent back by the rovers and orbiters is unmatched in the history of space exploration. Hundreds of thousands of detailed photographs are now stored in NASA databases, and new photos are transmitted every day.

We have so much data that it's actually hard to process it all. (Dr. Jeff Norris (2010), NASA Jet Propulsion Laboratory)[29]

The goal is to let the public participate in exploration, making contributions to data processing and analysis. It also provides a platform that lets developers collaborate with NASA on solutions that can help scientists analyse vast amounts of information to understand the universe and support future space exploration. The site was built using a variety of technologies, including the cloud-based Windows Azure platform, and Windows Azure Marketplace DataMarket—a service that lets developers and organisations create and consume applications and content on the Azure platform.

The 'Be A Martian' site has successfully demonstrated how Web technology can help an organisation engage with a large, dispersed group of users to view graphically rich content and participate in activities that involve massive amounts of data. Using the Azure DataMarket technology and leveraging Microsoft's cloud capacity, NASA created its experimental "Pathfinder Innovation Contest", which is designed to harness a global pool of programming and design talent to foster more citizen science contributions to Mars exploration.

---

[28]Microsoft website see www.news.microsoft.com/cloud/index.html retrieved on 3rd March 2015.

[29]Microsoft (2010) see www.microsoft.com/casestudies/Microsoft-Azure/Naspers-Pty-Ltd/New-NASA-Web-Site-Engages-Citizens-to-Help-Explore-Mars/4000008289 retrieved on 19th January 2015.

Amazon Web Services (AWS)

Previously, large data sets such as the mapping of the human genome required hours or days to locate, download, customise, and analyse. Now, anyone can access these data sets and analyse them using, for example, Amazon Elastic Compute Cloud (EC2) instances. Amazon EC2 is a web service that provides resizable compute capacity in the cloud. It is designed to make web-scale cloud computing easier for developers. By hosting this important data where it can be quickly and easily processed with elastic computing resources, AWS wants to enable more innovation, at a faster pace.

AWS hosts a variety of public data sets that anyone can access for free. One example of these public data sets, NASA NEX (NASA Earth Exchange), is a collection of Earth science data sets maintained by NASA, including climate change projections and satellite images of the Earth's surface. In 2013 NASA signed an agreement with AWS to deliver NASA NEX satellite data in order *"to grow an ecosystem of researchers and developers"*.[30] Previously, it had been logistically difficult for researchers to gain easy access to earth science data due to its dynamic nature and immense size (tens of terabytes). Limitations on download bandwidth, local storage, and on-premises processing power made in-house processing impractical. Through AWS, NASA is able to leverage the existing investment already made into the platform.

NASA NEX is a collaboration and analytical platform that combines state-of-the-art supercomputing, Earth system modelling, workflow management and NASA remote-sensing data. Through NEX, users can explore and analyse large Earth science data sets, run and share modelling algorithms, collaborate on new or existing projects and exchange workflows and results within and among other science communities. AWS is making the NASA NEX data available to the community free of charge.

> *We are excited to grow an ecosystem of researchers and developers who can help us solve important environmental research problems. Our goal is that people can easily gain access to and use a multitude of data analysis services quickly through AWS to add knowledge and open source tools for others' benefit.*[31] (Rama Nemani (2013), principal scientist for the NEX project at Ames)

> *Together, NASA and AWS are delivering faster time to science and taking the complexity out of accessing this important climate data.*[32] (Jamie Kinney (2013), AWS senior manager for scientific computing)

Scientists, developers, and other technologists from many different industries are taking advantage of AWS to perform big data analytics and meet the challenges of the increasing volume, variety, and velocity of digital information.

---

[30]NASA (2013) see www.nasa.gov/press/2013/november/nasa-brings-earth-science-big-data-to-the-cloud-with-amazon-web-services/#.VLK4KCusWSo retrieved on 19th January 2015.

[31]Ibid.

[32]Ibid.

Google Cloud Platform

Google's Cloud Platform is a set of modular cloud-based services that allow the user to create anything from simple websites to complex applications. Google's vast physical infrastructure enables it to build, organise, and operate a large network of servers and fibre-optic cables. Developers are therefore building on the same infrastructure that allows Google to return billions of search results in milliseconds. Google integrates with familiar development tools and provides API client libraries. Using Google's existing APIs and services can quickly enable a wide range of functionality for a developer's application, including the geo-services application, Google Maps.

Geo-services global revenues, made up of satellite navigation, satellite imagery, electronic maps and location-based search, are estimated at €125 billion–€225 billion per year.[33] There is therefore an increasing number of software developers offering location-based services (LBS) to consumers. Google Maps APIs give developers several ways of embedding Google Maps into web pages, and allows for either simple use or extensive customisation. When Google Maps was introduced, a programmer named Paul Rademacher introduced the first Google Maps mash-up (using content for an application from more than one source), HousingMaps.com, taking data from another Internet site, Craigslist.org, and creating an application that put Craigslist apartment and home listings onto a Google Map. Google hired Rademacher, and soon offered an API that made it easier for anyone to do what he did. Competitors, who had mapping APIs but locked them up behind tightly controlled corporate developer programmes, failed to seize the opportunity. Before long Google Maps had become an integral part of every web developer's toolkit. This is unlikely to have been possible without the existing investment in cloud infrastructure and systems underlying the application. Today, Google Maps accounts for nearly 90% of all mapping mash-ups, versus only a few percent each for MapQuest, Yahoo! and Microsoft, even though these companies had a head start in web mapping.[34]

Conclusion

Increasing value is created as data progresses through the value chain. Government agencies are working with private industry to stimulate innovative applications from the data. More data is being generated than ever before through the Internet

---

[33]Oxera (2013) *What is the economic impact of Geo-services, prepared for Google.*

[34]Lathrup, D. and Ruma, L. (2010) *Open Government: Collaboration, Transparency, and Participation in Practice*, O'Reilly Media.

of Things, information layers and diversified sources. An increased number of public and private (including new commercial players) EO satellite missions are contributing to an upsurge in EO data captured, with the intention of distributing this data to enable the creation of innovative applications. The cloud computing revolution continues to put downward pressure on the cost of processing big data, enabling larger datasets to be used. Turning the raw data into knowledge and actionable insights is where value is created. APIs, algorithms and open thematic platforms aim to extract value from raw data and turn pixels into insights. Democratising data use will engage a wider user community, therefore increasing the likelihood of stimulating new, disruptive innovations. To unlock the economic potential of this data, agencies are creating products that are responsive to user needs and implementing free and open data policies, including ESA's Copernicus Programme.

References

Council of the European Union (2015) Draft council conclusions on single market policy. Council of the European Union, Brussels

Meglio D et al (2014) CERN openlab whitepaper on future IT challenges in scientific research. Zenodo, Geneva

Digital Catapult (2014). https://digital.catapult.org.uk/. Accessed 2015

EARSC (2015) A survey into the state and health of the European EO services industry. EARSC, Brussels

Financial Times (2015). www.ft.com/cms/s/0/7805f624-a08b-11e4-8ad8-00144feab7de.html?siteedition=uk#axzz3PRjcpl00. Accessed 2015

The Economist (2013) A new goldmine; open data. www.economist.com/news/business/21578084-making-official-data-public-could-spur-lots-innovation-new-goldmine. Accessed 8 Feb 2015

Woodside Capital Partners (2014) OpenStack: is this the future of cloud computing? www.woodsidecap.com/wp-content/uploads/2014/11/WCP-OpenStack-Report_FINALa.pdf. Accessed 21 Nov 2014

8

The Open Science Commons for the European Research Area

Tiziana Ferrari, Diego Scardaci, and Sergio Andreozzi

Abstract Nowadays, research practice in all scientific disciplines is increasingly, and in many cases exclusively, data driven. Knowledge of how to use tools to manipulate research data, and the availability of e-Infrastructures to support them for data storage, processing, analysis and preservation, is fundamental. In parallel, new types of communities are forming around interests in digital tools, computing facilities and data repositories. By making infrastructure services, community engagement and training inseparable, existing communities can be empowered by new ways of doing research, and new communities can be created around tools and data. Europe is ideally positioned to become a world leader as provider of research data for the benefit of research communities and the wider economy and society. Europe would benefit from an integrated infrastructure where data and computing services for big data can be easily shared and reused. This is particularly challenging in EO given the volumes and variety of the data that make scalable access difficult, if not impossible, to individual researchers and small groups (i.e. to the so-called long tail of science). To overcome this limitation, as part of the European Commission Digital Single Market strategy, the European Open Science Cloud (EOSC) initiative was launched in April 2016, with the final aim to realise the European Research Area (ERA) and raise research to the next level. It promotes not only scientific excellence and data reuse, but also job growth and increased competitiveness in Europe, and results in Europe-wide cost efficiencies in scientific infrastructure through the promotion of interoperability on an unprecedented scale. This chapter analyses existing barriers to achieve this aim and proposes the Open Science Commons as the fundamental principles to create an EOSC able to offer an integrated infrastructure for the depositing, sharing and reuse of big data, including Earth Observation (EO) data, leveraging and enhancing the current e-Infrastructure landscape, through standardization, interoperability, policy and governance. Finally, it is shown how an EOSC built on e-Infrastructures can improve the discovery, retrieval and processing

T. Ferrari • S. Andreozzi
EGI Foundation, Amsterdam, The Netherlands
e-mail: Tiziana.Ferrari@egi.eu; sergio.andreozzi@egi.eu

D. Scardaci (✉)
EGI Foundation & INFN Catania Division, Amsterdam, The Netherlands
e-mail: diego.scardaci@egi.eu

capabilities of EO data, offering virtualised access to geographically distributed data and the computing necessary to manipulate and manage large volumes. Well-established e-Infrastructure services could provide a set of reusable components to accelerate the development of exploitation platforms for satellite data solving common problems, such as user authentication and authorisation, monitoring or accounting.

Creating the European Research Area: The "Open" Approach

The European Research Area (ERA) was endorsed by the European Council in 2000 (European Commission 2014a) as a way to build *"a unified research area open to the world based on the Internal Market, in which researchers, scientific knowledge and technology circulate freely and through which the Union and its Member States strengthen their scientific and technological bases, their competitiveness and their capacity to collectively address grand challenges"* (European Commission 2012).

Although several actions for the ERA implementation have been undertaken, such as the establishment of the European Strategy Forum on Research Infrastructures (ESFRI) (2017) or the development of an e-Infrastructure for connectivity, high performance, grid and cloud computing and data, the rapid growth of scientific data has highlighted the need for new methodologies.

In its Horizon 2020 consultation report on Open Infrastructures for Open Science, the European Commission concluded that *"open data e-Infrastructures increase scope, depth and economies of scale of the scientific enterprise. They are catalysts of new and unexpected solutions to emerge by global and multidisciplinary research. They bridge the gap between scientists and the citizen and are enablers of trust in the scientific process"* (Horizon 2020 consultation report n.d.), electing the *open* approach as a core aspect of the ERA.

This vision implies a European dimension beyond national and regional approaches, and an increase in capacities and capabilities. Research Infrastructures, including e-Infrastructures, are enabling instruments which provide (European Commission 2017) *"facilities, resources and services that are used by the research communities to conduct research and foster innovation in their fields. They include: major scientific equipment (or sets of instruments), knowledge-based resources such as collections, archives and scientific data, e-infrastructures, such as data and computing systems and communication networks and any other tools that are essential to achieve excellence in research and innovation."*

Problems to Solve

Implementing such vision requires overcoming various barriers.

Lack and/or Incomplete Roadmaps for Research- and e-Infrastructures

The establishment of roadmaps at a national and European level is progressing at different speeds. Policies of access to the European research system are not homogeneous, services are not always available to researchers, and knowledge transfer is not yet a strategic mission of many public organisations. This represents a risk to the coherence of the European-scale initiatives. For example, with inconsistency in access policies, countries may find their national research communities excluded from crucial European-wide services. Shared understanding of the strategic objectives, the alignment of national strategies and proceeding faster in the development of the national policies are recognised by the European Council as necessary steps: *"the Member States should accelerate national reforms, where necessary, to boost the EU's potential in research, development and innovation"* (European Commission 2014b).

The lack of alignment in policies hinders the sustainability of e-Infrastructures of European scale. A consensus solution would offer them the certainty of a long-term commitment to allow for the construction of technical infrastructures and high-quality services.

Fragmented Solutions and Policies for Access to Data and Knowledge

Access to research data and existing bodies of knowledge has moved towards openness, but has not yet achieved it. Efforts to support Open Access are hugely beneficial, but the scientific publishing sector is still adjusting to the evolution towards openness. The resulting inconsistency restricts exploitation of research results.

Insufficient Cooperation Between Public and Private Sector

There is a general agreement amongst research communities and e-Infrastructure providers that working with the private sector is desirable. However, the mechanisms and models of engagement are not yet well understood, due to a range of technical and cultural factors. Steps such as SME instruments in Horizon 2020 have been broadly appreciated, but many challenges remain, due to different local strategies, restrictions in policies and insufficient European coordination of existing efforts.

Lack of National and European Organization Between All Stakeholders

While the vision of the e-Infrastructure commons (e-Infrastructure Reflection Group 2017a) has been embraced by many groups, the landscape remains fragmented and includes too many narrowly focussed services based on closed platforms that limit the portability of data, applications and knowledge. In addition, we are still missing a common body of knowledge and a coordinated broad programme for knowledge transfer, also including the private sector and the single researchers. This results in a barrier to entry for the emerging research infrastructures, skills and professions.

Today, services are provided by a broad range of sector-based, national and pan-European providers. Technical interoperability and service integration and management are becoming increasing important to ensure the support of the entire research lifecycle. Nevertheless return on investment for national and European funds dedicated to service support, is still sub-optimal due duplicated efforts, limited sharing of technical solutions and some lack of coordination.

As digital science services such as e-Infrastructures move toward sustainable operating models, the need for coordination and coherence is rapidly increasing.

Many Providers Without a Single Market

Despite some efforts, we lack a single portfolio of services to provide a "backbone" of European ICT capabilities. The existing offering is fragmented, with different policies of access and different channels for engagement with the user community. One of the main policy issues that are hindering the uptake of services at national level is the lack of a business model and procurement framework that allows international research collaborations to rely on research e-Infrastructure services in the long term.

In addition, the lack of an established national Research- and e-Infrastructure component in some countries prevents the access even to the existing services. Such European-scale coherence must be achieved to make the ERA a reality. Insufficient competition in national research systems, barriers to pan-European cooperation and restricted circulation of and uneven access to scientific knowledge are also recognized as issues by the ERA implementation assessment.

The Open Science Commons

The Open Science Commons have been defined as an overarching policy designed to overcome the barriers preventing the implementation of the ERA. The Open Science Commons seek to encompass all the elements required for a functioning ERA: **research data, scientific instrumentation** (such as the Large Hadron Collider, the Copernicus satellites or Square Kilometre Array), **ICT services** (connectivity, computing, platforms and research-specific services such as portals), and **knowledge**. The Open Science Commons evolve from two preexisting and broadly accepted ideas:

- **Open Science** (also referred as Science 2.0 (European Commission, n.d.)) for the opening of knowledge creation and dissemination to a multitude of stakeholders, including society in general. Open Science supports multiple perspectives, from infrastructure-oriented views seeking to increase efficiency through better tools and services, to public-oriented views trying to ensure citizens have access to scientific knowledge (Fecher and Friesike 2014).

- A **Commons** to reinforce the need of sharing within a community in a way that allows non-discriminatory access, while ensuring adequate controls to avoid congestion or depletion when the capacity is limited (Frischmann 2013). The Commons concept is embedded in Open Science, which stresses cooperation to reduce barriers to collaboration, knowledge transfer and sharing of results. In the last decade, this has been driven by the digitalisation of the research process and by the globalisation of the scientific communities. Infrastructures often generate spill overs that result in large social gains and it is recognised that applying commons management principles maximises such benefits.

The Open Science Commons relies on **four pillars**, representing a wide range of groups, providers and community types:

- **Data**. The data that is the subject matter for research. It should be dealt with according to the principles of open access and open science, while maintaining trust and privacy for researchers.
- **e-Infrastructures**. The technology and technical services supporting researchers, building towards integrated services and interoperable infrastructures across Europe and the world.
- **Scientific instruments**. The equipment and collaborations that generate scientific data, from small-scale lab machines to global collaborations around massive facilities.
- **Knowledge**. The human networks, understanding and material capturing skills and experience required to carry out open science using the three other pillars.

Managing shared resources as a Commons maximises benefits for society. In the area of digital infrastructure, this has already demonstrated great benefits (the Internet itself is an example). Applying this principle to the Open Science process is expected to improve the stewardship from the funding agencies, in collaboration with the stakeholders, through mechanisms such as public consultations. This will increase the perception of shared ownership of the infrastructures. It will also create clear and non-discriminatory access rules together with the sense of shared ownership, which stimulates a higher level of participation, cooperation and social reciprocity.

The European Open Science Cloud

As part of the European Commission Digital Single Market strategy (European Commission 2015), the European Open Science Cloud (EOSC) initiative was officially launched in April 2016 by the European Commission. EOSC promotes not only scientific excellence and data reuse but also job growth and increased competitiveness in Europe, and drives Europe-wide cost efficiencies in scientific infrastructures through the promotion of interoperability on an unprecedented scale.

According to the first report of the High Level Expert Group on the European Open Science Cloud (EOSC) (European Commission 2016) appointed by the European Commission, EOSC has been defined as a support environment for Open Science aiming to *"accelerate the transition to more effective Open Science and Open Innovation in a Digital Single Market by removing the technical, legislative and human barriers to the re-use of research data and tools, and by supporting access to services, systems and the flow of data across disciplinary, social and geographical borders"*. Indeed, the term "cloud" has been interpreted as *"a metaphor to help convey the idea of seamlessness and a commons"*.

As guiding principles, the experts underlined how:

1. The EOSC must integrate with other e-Infrastructures and initiatives in the world, implementing a lightweight interconnected system of services and data, which follows the federated model.
2. Open refers to the accessibility of services and data under proper non-discriminatory policies (*"not all data and tools can be open"* and *"free data and services do not exist"*).
3. EOSC should include all scientific disciplines.
4. The term cloud should not refer to ICT infrastructure but to a commons of data, software, standards, expertise and a policy framework relevant to data-driven science and innovation.

In the expert vision, the EOSC will be an accessible infrastructure for modern research and innovation implementing an internet of Findable Accessible Interoperable and Reusable (FAIR) data and services (Wilkinson et al. 2016). It should be based on standards, best practices and infrastructures, completed with adequate human expertise. The FAIR principles have to be supported, with a particular attention to the reuse of the open and sensitive data. The data needs to be sustained with a plethora of elements (standard, tools, processing pipelines, protocols) that make possible and simple their reuse, enabling a data driven knowledge discovery and innovation. Furthermore, the data science profession needs to be fostered to guarantee a professional data management and the long-term data stewardship.

In Europe, domain-specific European Research Infrastructures and cross-domain ICT e-Infrastructures, as well as other disciplinary and cross-disciplinary collaborations and services are already well established. These can be considered the ground for EOSC. However, the realisation of *"the ambition of increased seamless access, reliable re-use of data and other digital research objects, and of the collaboration across different services and infrastructures"* (that guarantees non-discriminatory access and reuse of data to both the public and private sector), requires further enhancements on this landscape with the aim to turn the *"ever increasing amounts of data [. . .] into knowledge as renewable, sustainable fuel for innovation in turn to meet global challenges*". The EOSC is the instrument defined by the European Commission to foster such evolution towards the realisation of the so-called Open Science.

This idea highlights the strong relationship between the implementation of the ERA through the Open Science, the Open Science Commons and the EOSC. In such context, the High Level Expert Group designed by the EC reported a list of key Open Science "trends" that should be taken into account in the EOSC design. They cover several aspects such as new modes of scholarly communications (e.g. applications, software pipelines and the data itself), new incentives to stimulate data publishing and tool sharing, fostering the emerging data science profession, the cross-disciplinary collaboration, support for innovative SMEs, the creation of an ecosystem, methodology and tools to reproduce of current published research, etc.

Open Science Commons for the EOSC

In this section, we propose the Open Science Commons as a possible approach to the implementation of the infrastructure and governance pillars of the EOSC, leveraging and enhancing the current Research Infrastructure landscape, through standardization, interoperability, policy and governance. The section discusses the current state of play and blockers for implementing an Open Science Cloud, and explains how the approach could strategically advance the competitiveness of research in Europe by providing research data and community-specific tools as services through a platform that supports the participatory principle of Open Science.

EOSC Architecture and Services

The Open Science Commons infrastructure comprises research data, processing services, applications, virtual laboratories and tools, relying on existing *federated* data and storage facilities from local, regional and international infrastructures, which can be organized as a federation of hubs, where each hub is a node of the EOSC providing certain capabilities in a standard and interoperable manner (Fig. 1). The cloud hub does not duplicate the data services of the reference institutional and disciplinary repositories, but rather make these accessible in an environment that can enrich the data itself with supplementary added-value services and can provide scalable access where necessary by collocating computing and data.

In the proposed approach the Open Science Commons can be implemented as a federation of cloud hubs (in Europe and beyond), based on the cloud service provisioning paradigm. The cloud hub will provide various capabilities in a federated, integrated way: a virtual space providing data, tools, applications and processing, with the hubs interconnected by a mesh of high-bandwidth links to

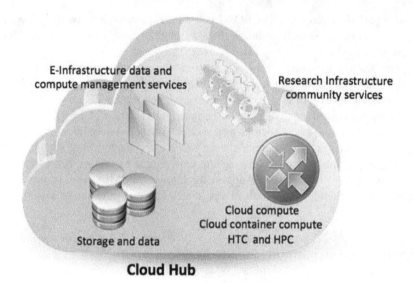

Fig. 1 Examples of functionalities delivered by a cloud hub of the EOSC

ensure efficient virtual access to public and managed access research data, which is provided as a service by the hub (DaaS). Within the cloud hub, the data provider always retains access control to data.

Being based on virtualization, clouds facilitate sharing, reuse and the combined offer of data and tools. Cloud federations enable "local hosting" and "control sharing" capabilities to respect ownership and allow accessibility for distributed communities of users. In addition, federation allows the implementation of hybrid models where private, community and public providers can contribute data and services in multi-supply environment.

Furthermore, the federation of hubs provides a multi-level governance model where different governing bodies of the Commons can coexist. This governance model meets the needs of European policies, regulations, restrictions and business models. By allowing distributed access to data, relocation into centralized repositories is no longer necessary. This greatly simplifies the integration of data and tools from multiple domains and regions. When expertise about how to use specific research domain data and tools is accumulated within the same research community, then the community becomes an ideal incubator for a hub and can contribute to the implementation of the EOSC federated infrastructure.

Realizing a Federated Approach to Research Data

The EOSC needs to aggregate offer and demand by exposing its assets via a marketplace to make research data, the related tools and knowledge discoverable, accessible and reusable. The marketplace would federate existing research data sets that are provided by data preservation organizations that can ensure compliance to a set of defined quality standards. The EOSC Marketplace should be open to any data provider that can ensure compliance to international data standards and best practices, as well as to European data regulations.

The Marketplace should be open for access to any research community that is willing to become a data provider. Through the marketplace, datasets and the associated metadata are discoverable. The marketplace provides information about intellectual property rights and access policies for reuse for research and commercial purposes when allowed.

Offering of Scalable Access to and Analysis of Research Data for Reuse

Making data findable is not sufficient. Local download of large volumes of data can be a huge barrier for downstream efficient analysis. EOSC should provide distributed data mirroring and caching capabilities based on federated cloud storage, where research data can be temporarily stored for scalable access in agreement with the data providers, and processed via integrated computing platforms.

This capability is not a duplication of existing data infrastructures, but rather provides efficient access to big data that is produced worldwide. The governance of the service would require an organization acting as a broker towards the data providers worldwide for the procurement of Data as a Service to the whole ERA.

A premium access could be also offered implementing a federation of large cloud hubs connected by a broadband network infrastructure for efficient replication. The network of Tier-1 hubs would be complemented by a network of disciplinary Tier-2 hubs, providing complimentary access to discipline-specific datasets. The cloud hub federation would be complemented by co-located services offering high throughput and high performance cloud computing.

Integrating (Shared) Tools and Applications

Knowledge cannot be extracted from data without the availability of specialized tools and applications (e.g. text mining). The EOSC would provide a library of community-specific applications and tools. This community platform should be open for publishing to any researcher. For greater specialization, the EOSC should provide PaaS and SaaS services that are community-specific and that could be dynamically deployed with a focus on single researchers or small research groups. These are provided in the form of managed services by the Research Infrastructures. By increasingly sharing models and modelling tools, researchers and research communities can capture the steps of the digital research processes they carry out for excellent science. With suitable abstractions and robust provenance capabilities, such models and tools would enable the repeatability, and therefore the incremental improvement of research practices and processes within and across research teams.

Provisioning of Services for Depositing Data for Resource-Bound Users

Through virtual access, the EOSC will federate e-Infrastructures to provide services for the long tail of science, citizen scientists, the general public and other stakeholders that cannot benefit from those at institutional and/or national level, but supports open research data.

EOSC Service Integration and Management

As the EOSC will involve multiple suppliers, the EOSC requires a service integration and management approach to managing suppliers and integrating them to provide a single customer-facing interface. The approach allows integrating interdependent services from various internal and external service providers into end-to-end services.

EOSC service integration and management plays different roles. It allows the end-to-end composition of services, the alignment of scope, value, service catalogue entries and their specifications across EOSC providers, the management of relationship and collaboration between the providers, and the EOSC standardisation guidelines. By doing so, the EOSC governance becomes accountable for the integrated services that are delivered and for the central point of control between demand and supply.

The EOSC service integration and management system includes the entirety of activities performed by service providers to plan, deliver, operate and control services offered to customers. These activities are directed by policies and need to be structured and organised by processes and procedures.

The processes include the management of Business Development and Stakeholders, Service Order, Service Portfolio, Service Level and Reporting, Customer Relationship, Supplier/Federation Member Relationship, Capacity, Service Availability and Continuity, Incident and Service Request, Problem, Configuration and Change, and finally Release and Deployment (Fig. 2).

EOSC Governance

Cloud hub services could be provided in a coordinated fashion by multiple stakeholders, including research communities, research infrastructures and e-Infrastructures. In this case, a *federator* role needs to be established to ensure services are provided in an integrated way according to a single community-defined lightweight service integration and management framework that defines a corpus of policies (the so-called "rules of engagement"), processes and tools for aggregating demand and supply from local, national and regional facilities.

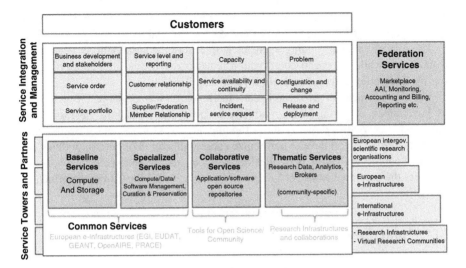

Fig. 2 The EOSC infrastructure includes a service catalogue (*bottom layer*) and a service integration and management services and activities (*top layer*) that ensure the delivery of services in a multi-supply environment. The federation services (*right*) support the service integration and management system, and enable the federation of distributed facilities. EOSC services (the "towers"), aggregated in cloud hubs and in the EOSC will be delivered by multiple suppliers: European intergovernmental organizations, European/international Research e-Infrastructures, and virtual research communities

The value proposition of the Open Science Commons is the combination of services and their integration layer, which altogether augments the existing data infrastructures and allows extraction of knowledge and the generation of innovation from research data. In such vision, open standards are considered as enablers of the Commons.

The realization of an EOSC must avoid duplication of provisioning of ICT services at national and European level and ensure efficient provisioning. National EOSC facilities are primarily financially supported by the European Member States. The role of the European Commission would be to ensure the persistency of the services that allow the national cloud hubs to operate as a federation, and to ensure the coordinated procurement, service provisioning and data brokering according to the requirements of the RIs. This would allow aggregation of demand across Europe, coordinated delivery and the development of economies of scale.

This would increase the coordination between Research Infrastructures, e-Infrastructures and data providers in matters concerning ICT provisioning. With European coordination, an economically efficient system of tools can be developed, which can accelerate the flourishing of multidisciplinary science, open science and a sustainable system of integrated Commons.

The EOSC can be organised as an integration of existing e-Infrastructures with overarching governance and common agreed services. The EOSC can be created based on both publicly funded and commercial providers as long as they are all based on open standards and remove the risks for artificial lock-in.

The role of national funding agencies is to ensure the sustainability of national cloud hubs, while European coordination is necessary to focus on supporting the service integration and management layer of the EOSC.

The development of a research data marketplace will promote the definition of governance involving data providers, archiving organizations, infrastructure providers, knowledge organizations, funding agencies, users and citizens. This EOSC governance will have to harmonize access, allocation of quotas, policies and acting as conflict resolution body. Besides being inclusive, the governance will need to reflect the federated nature of Europe and be inspired by the Commons principle.

EOSC and the e-Infrastructure Commons

The e-Infrastructure Reflection Group (e-IRG) (2017b) is a strategic body facilitating integration in the area of European e-Infrastructures and connected services, within and between member states, at the European level and globally. The mission of e-IRG is to support both coherent, innovative and strategic European e-Infrastructure policymaking and the development of convergent and sustainable e-Infrastructure services.

e-IRG has recently published a new version of its roadmap document. The e-IRG Roadmap 2016 is taking up the e-Infrastructure Commons concept, introduced in the previous version of the roadmap (2012) (e-Infrastructure Reflection Group 2012). The document intends to define a clear route on how to evolve the European e-Infrastructure system further. The new issue aims to turn the vision of the e-Infrastructure Commons, one of the pillars of the Open Science Commons, into reality by 2020 and has recommendations to all stakeholders to progress on the way towards the e-Infrastructure Commons. The key recommendations in the roadmap document encourage the research infrastructures and research communities to elaborate on and drive their e-infrastructure needs.

A concrete way towards the e-Infrastructure Commons, loosely integrating the different types of e-Infrastructures, is to use a marketplace with a proper governance including a representation of the users as a single point of access to all e-Infrastructure services and tools. The marketplace will act as a one stop-shop for EU researchers, i.e. a place where all e-Infrastructure services are accessible together, either directly or redirected elsewhere. The marketplace can make use of several technologies and services, such as cloud technologies, a searchable service catalogue and a common authentication/authorisation scheme. In this way, a standardised and single point of access to services will be achieved, without promoting monopolies, nor a single integrated provider. This has proven to be very difficult across different e-Infrastructure components. On the contrary, it will be open to new actors, encouraging cooperation, competition and innovation.

The e-IRG roadmap 2016 clearly identifies e-Infrastructures as the *seeds* to implement the EOSC and defines a clear roadmap towards 2020 to deal with the challenges described in the previous section.

The role of the e-Infrastructures to establish the EOSC is also described in the position paper *European Open Science Cloud for Research* written in collaboration by five leading European initiatives, EUDAT (2017), LIBER (2017), OpenAIRE (2017), EGI (2017a) and GÉANT (2017). In that paper, the relevance of the Open Science Commons for the EOSC has been underlined as *a key driver, not only of scientific progress, but also of economic and societal innovation.* The EOSC has been presented as a vehicle *to foster an open, collaborative platform for the management, analysis, sharing, reuse and preservation of research data on which innovative services can be developed and delivered.* Furthermore, the paper considers fundamental leveraging on the scientific infrastructures already available in Europe as outcome of decades of public investments. The EOSC could be developed by connecting the existing national and international infrastructures and services. Indeed, many of the resources and services needed for the Open Science Cloud already exist, but while technical challenges remain, *most of the barriers are ones of policy and concern funding, lack of interoperability, access policies and coordinated provisioning.* Then, the EOSC should address these issues and enhance both the service portfolio and the amount of available resources.

The EGI Blueprint

The key role of the e-Infrastructures on creating the EOSC has been clearly introduced in the previous section. Now, as example, key services, platforms and tools of one of the main European e-Infrastructure, EGI, will be presented highlighting how these could be the ground for implementing the Open Science Commons and, then, the EOSC. In particular, already available features that could facilitate the building of the Cloud Hub model for EOSC previously defined will be underlined.

EGI, advanced computing for research, is a federated e-Infrastructure set up to provide advanced computing services for research and innovation. The EGI e-infrastructure is primarily publicly funded and comprises over 300 data centres and cloud providers spread across Europe and worldwide. EGI offers a wide range of services for compute, storage, data and support (EGI 2017b) and provides access to over 700,000 logical CPUs and 500 PB of disk and tape storage. Its principles are based on the Open Science Commons and its mission is creating and delivering open solutions for science and research infrastructures by federating digital capabilities, resources and expertise between communities and across national boundaries.

The EGI architecture is organised in platforms (Fig. 3):

- Core Infrastructure Platform, to operate and manage a distributed infrastructure;
- Cloud Infrastructure Platform, to operate a federated cloud-based infrastructure;

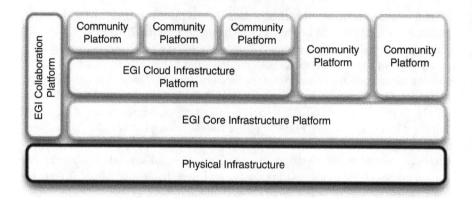

Fig. 3 EGI platform architecture

- Open Data platform, to provide easy access to large and distributed datasets;
- Collaboration Platform, for information exchange and community coordination, and
- Community Platforms, tailored service portfolios customised for specific scientific communities.

The platform architecture allows any type and any number of community platforms to co-exist on the physical infrastructure.

In the remaining part of the section, the platforms that could provide functionalities useful to implement the EOSC are shortly introduced.

Core Infrastructure Platform

The **EGI Core Infrastructure** provides all the necessary operational tools and processes to operate and manage a large distributed infrastructure guaranteeing standard operation of heterogeneous infrastructures from multiple independent providers. This also includes:

- The **Authentication and Authorisation infrastructure** for homogeneous authentication and authorisation across the whole federation.
- A **Service registry** for configuration management of federated services.
- **Monitoring** tools, performing service availability monitoring and reporting of the distributed service end-points.
- **Accounting** for collecting, and displaying usage information.
- **Information discovery** about capabilities and services available in the federation.
- **Virtual Machine image catalogue** and distribution: allows researchers to share their virtual appliances for deployment in a cloud federation.

The federated environment of European e-Infrastructures, implemented in EGI through the Core Infrastructure platform, is a key enabler for distributed management and processing of big data and a fundamental baseline to implement the service integration and management system of the EOSC.

Collaborative Platform

It provides IT Infrastructure and services that facilitate collaboration between research communities. Its two main components are the Marketplace and the Application Database.

The Marketplace (EGI 2017c)

The marketplace has the ambition of becoming the platform where an ecosystem of EGI-related services, delivered by providers and partners, can be promoted, discovered, ordered, shared and accessed, including EGI offered services as well as discipline and community-specific tools and services enabled by EGI and/or provided by third parties under defined agreements.

The need of a Marketplace, making discoverable open research data and the related tools and knowledge, and acting as a one stop-shop for EU researchers, has been identified in the Service Hub model to build the EOSC and in the e-IRG roadmap 2016 introduced before. This should also act as a single point of access to all e-Infrastructure services and tools for all users. The EGI Marketplace could be seen as first test to implement such tool (to be extended both in term of features and coverage).

Application Database (AppDB) (EGI 2017d)

It is a tool that stores and provides information about:

* software solutions in the form of native software products and virtual appliances,
* the programmers and the scientists who are involved, and
* publications derived from the registered solutions.

Reusing software products registered in the AppDB, means that scientists and developers may find a solution that can be directly utilized on the infrastructure. In this way, scientists can spend less or even no time on developing and porting a software solution to the Distributed Computing Infrastructures (DCIs) and facilitate the reproducibility of experiments. AppDB, thus, aims to avoid duplication of effort across the DCI communities and to inspire scientists who are less familiar with DCI programming and usage. The service is open to every scientist interested in publishing and therefore sharing their software solution.

The AppDB can be considered as an example of a service providing a library of community-specific applications and tools that could enable the repeatability, and therefore the incremental improvement of research practices and processes within and across research teams.

The EGI Federated Cloud

EGI launched the production phase of a cloud federation to serve research communities in May 2014, the EGI Federated Cloud (EGI 2017e). It integrates community, private and/or public clouds into a scalable computing platform for data and/or compute-driven applications and services.

Its architecture is based on the concept of an abstract Cloud Management Framework (CMF) that supports a set of cloud interfaces to communities. Each resource centre of the infrastructure operates an instance of this CMF according to its own technology preferences and integrates it with the federation by interacting with the EGI Core Infrastructure platform.

This integration is performed by using public interfaces of the supported CMFs, thus minimising the impact on site operations. Providers are organised into realms exposing homogeneous interfaces and grouping resources dedicated to serve specific communities and/or platforms.

The EGI Federated Cloud is based on a hybrid model where private, community and public clouds can be integrated and already offers some of the facilities that a Service Hub should provide such as the virtualisation and the easy share and reuse of tools (Fig. 4).

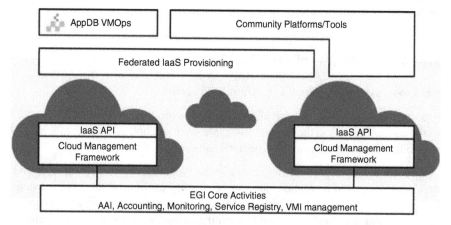

Fig. 4 EGI Federated Cloud architecture: each resource centre of the infrastructure operates an instance of a CMF according to its own technology preferences and integrates it with the federation by interacting with the EGI Core Infrastructure platform. Providers are organised in realms exposing homogeneous interfaces (Federated IaaS provisioning). Community platforms can exploit resources from one or more realms through such interfaces. The AppDB VMOps enables an automatic deployment of virtual appliances on all the resource centres supporting a specific community

The Data Hub and the Open Data Platform

The Data Hub provides easy and efficient access to large-scale datasets enabling sharing, discovering, and processing of data federated from different sources. The service offers a virtual access to files distributed across different types of storage and geographically distributed providers through homogenous and standard based interfaces (POSIX, CDMI, etc.).

The technology behind the Data Hub service is the Open Data Platform, implemented in the EGI-Engage project (2017), aiming at overcoming the technical barriers that are still faced to federated data on cloud across multiple storage providers. Its design emerged from the analysis of several user communities' requirements, including some of the major Research Infrastructures on the ESFRI roadmap, with focus on open data management. It allows the integration of various data repositories available in a distributed infrastructure, offering the capability to make data open, and link them to key open data catalogues following respective guidelines, such as the OpenAIRE (2017) open access infrastructure. The core enabling technology of ODP is Onedata (CYFRONET, n.d.), a data management solution that allows a seamless and optimised access to data spread over a distributed infrastructure (see Fig. 5).

The Data Hub service can help on implementing the Cloud Hub based architecture for EOSC dealing with the offering of scalable access to and

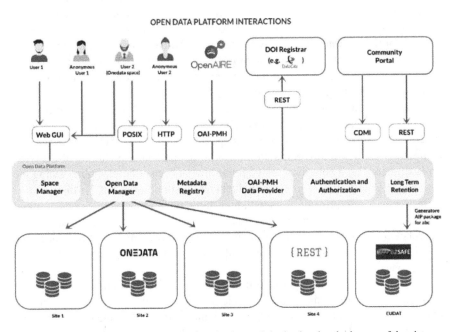

Fig. 5 Open Data Platform architecture, showing its modular backend at the *bottom* of the picture, and how it integrates with different data management systems in its distributed configuration at different data centres ("sites")

analysis of research data for reuse. Indeed, it can not only make data and metadata findable through its catalogues and metadata, but can also provide the distributed data mirroring that, integrated with the IaaS features offered by the Federated Cloud, allows an efficient and scalable processing of the datasets reducing or removing the needs of data movements.

The future evolution of the Data Hub foresees the introduction of a smart caching mechanism that could cleverly move portions of dataset before the users' request dramatically decreasing time to access. These caching mechanisms will take into account factors like the number and type of data analysis applications that are currently running in the infrastructure, the popularity of datasets and the catalogue of recently generated data. Such factors will drive the geo-replication of the data, implementing a network of distributed data hubs that minimizes the need of data transfers, by coupling storage and computing resources and caching heuristics.

EO Data Exploitation via the e-Infrastructures and EOSC

The new generation of Earth Observation (EO) satellites from the Sentinel missions that ESA developed for the Copernicus programme, is generating large amounts of data that are not easily integrated into processing chains outside the Copernicus ground segment. Very often, public and private institutions aiming to deliver end-user services based on EO data do not possess the computing power, the storage capacity or the software technology to cope with these new data flows. Handling an increasing volume of EO data is one of the main challenges for the community and hybrid clouds, coupled with big data management solutions and applications, are seen as a potentially efficient solution.

E-Infrastructures can improve the discovery, retrieval and processing capabilities of Copernicus data through their capabilities for the management of big data. Indeed, they offer virtualised access to geographically distributed data and the computing necessary to manipulate it, and to manage large volumes of different types of data. These solutions can be reused by any scientific discipline facing the challenge of integrating large datasets including earth observation. Furthermore, e-Infrastructures also provide users with high-throughput access to data sets, without the necessity of prestaging data on local storage, thus enabling full support for hybrid cloud scenarios. This scenario gives users the freedom to mix computational and data storage resources from various infrastructure providers, including their own private ones.

In addition, well-established e-Infrastructure services complete the set of reusable components that could accelerate the development of exploitation platforms for satellite data solving common problems, such as the user authentication and authorisation, the monitoring, the accounting, etc.

This already promising scenario needs enhancements to overcome the barriers previously described and, the implementation of the EOSC vision introduced in this chapter draws a clear roadmap to enhance the e-Infrastructure services in the next years. However, the current maturity of the e-Infrastructure services already allows their use for the EO data exploitation. This is detailed in the next section.

Infrastructure Services for EO Data Exploitation

The design of appropriate solutions for managing and exploiting large amount of EO Datasets needs expertise from different sources, such as:

- Data consumers, to help defining and designing real added-value services to be integrated with the existing EO Exploitation Platforms;
- ICT experts and platforms operators, with advanced knowledge on EO systems, to develop and provide hosting platforms for these services and to offer general solutions;
- E-Infrastructures, to supply the computing and storage resources needed for data access and exploitation and provide the tools to manage the datasets in a distributed environment

The joint expertise from these three sectors can allow the creation of an integrated environment for fast development and prototyping of services for exploitation platforms and scientific applications.

E-Infrastructures are the foundation of such an environment and could become an attraction pole to merge these sectors and enable such potential. By bringing these individual groups together, the fostering of new knowledge through the sharing of specialised expertise will be tremendously accelerated. This will ultimately improve innovation capacity leading to fast development and prototyping of services able to deal with so extensive sources of information through the integration of e-Infrastructure services, EO exploitation platforms and scientific applications. E-Infrastructure services could also evolve according to the specific needs identified by both data consumers and EO platform operators. This will facilitate the exploitation of EO data with the final aim to facilitate the creation of applications to address societal challenges, enabling policymakers and authorities, including environmental agencies to develop long-term strategies as well as react efficiently to sudden crisis situations.

Furthermore, the adoption of the e-Infrastructures for EO Data exploitation will facilitate the market uptake of the satellite data, including those coming from the Copernicus programme, and will contribute to the creation of a European solution for exploiting Earth Observation data, fostering EO services within the public and private sectors. About the latter, widely diverse and small companies throughout Europe could benefit from the easier data access to develop new products and services.

The following image shows the role of the e-Infrastructure services for the EO Data exploitation in the European Scenario (Fig. 6).

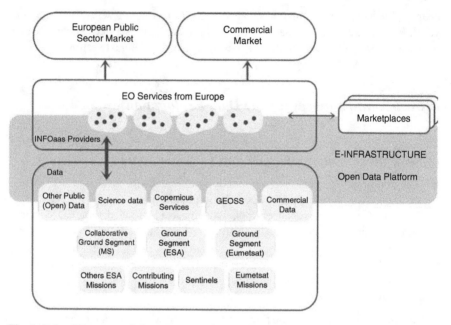

Fig. 6 Role of European e-Infrastructure services for EO data exploitation

An Example: e-Infrastructure Services to Implement the ESA Generic Exploitation Platform Open Architecture

As example, this section shows how the e-Infrastructure services could facilitate the implementation of a EO exploitation platform based on the generic Exploitation Platform Open Architecture (ESA 2016) defined by ESA with the aim to harmonize the architecture of such platforms (in particular the ESA Thematic Exploitation Platforms—TEPs) and make the adoption of shared solutions easier.

ESA Exploitation Platform Open Architecture

ESA defined the concept of a General Exploitation Platform Architecture (Fig. 7) to harmonize the different Exploitation Platforms and maximize the reuse of technology and components, thus reducing the cost required to develop, maintain and operate them. This architecture includes a set of common components, to be reused by the single platforms and tailored to the particular platform needs.

- **User access portal**: it is the interface for the final users and the system operators. It includes services used by the other components, such as authentication, accounting, monitoring and collaboration tools and implements a *single access portal, collaboration tools, documentation and support tools, services mar-*

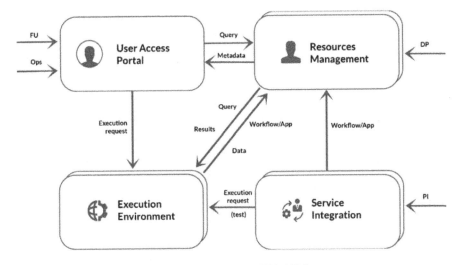

Fig. 7 ESA Exploitation Platform Open Architecture (ESA 2016)

ketplace and *operator interface* functionalities. It interacts with the Resource Management component for getting information about the data, application and workflow resources required for a given processing service and with the Execution Environment for executing processing services.

- **Resource management**: handles the resources available in the platform and implements *data discovery*, *data management* and *processing services management* functionalities. It provides the User Access Portal and the Execution environment components with interfaces to:

  - Perform query to retrieve data products, data results, application or workflow packages and the related metadata,
  - Publish new data or applications provided by a data provider or a scientific user.

- **Service Integration**: provides the programmatic interface with a framework to integrate applications, algorithms and/or software into the platform as a new processing service. It implements the required *development and integration system* functionality. It can send requests to execute an application to the Execution Environment for testing purposes.
- **Execution Environment**: provides the platform with an environment to run processing services. It implements both the *on-demand processing* and *massive processing* functionalities. It receives the application and data required to run the processing from the Resource Management component. It can also perform queries to the Resource Management component to identify the resources required by the processing that are not directly specified in the execution request.

Infrastructure Services to Implement the ESA Exploitation Platform Open Architecture

The following table shows the relations between each of the ESA Exploitation Platform components and the EGI services highlighting the features of such services that support the implementation of the TEPs functionalities.

TEPs macro-components	EGI services	Functionalities
User access portal	EGI core platform services: AAI infrastructure, accounting and monitoring.	Single access portal: AAI infrastructure Services: accounting and monitoring
Resource management	EGI Data Hub & Open Data platform	Data discovery: query to the Data Hub metadata catalogue Data management: registration and setup of EO data products into the Data Hub metadata catalogue and related pre-staging through the smart cache
Service Integration	N.A.	N.A.
Execution Environment	EGI Federated Cloud: OCCI and Open Stack standard interfaces Open Data platform	On-demand and massive processing: (1) standard OCCI and Open Stack interfaces to access a public cloud to run to run workflows, apps or bulk processing, (2) list of computing resources close to the pre-staged data

This mapping demonstrates that several TEP functionalities, described in the ESA Exploitation Platform open architecture, could already benefit from the EGI services. EGI already validated this statement having supported the integration of the Geohazard Exploitation Platform (GEP) (2017) within its FedCloud in the last years. Such result could be advertised to other ESA TEPs that could also benefit from the EGI services. This fits very well with the ESA vision of reusing technology and components to reduce the cost. The integration of the Hydrology TEP is also already planned.

An Example: Integration of the Geohazard Exploitation Platform Within the EGI Infrastructure

The Geohazard Exploitation Platform has been developed by Terradue on behalf of ESA. Terradue has integrated such platform with the EGI infrastructure developing the needed extensions in their workflows to use the EGI Federated Cloud interfaces and tested in several resource centres belonging to the EGI infrastructures.

Terradue has extended their workflows to be compatible with the following EGI components:

- AAI;
- Federated Cloud: Implemented an OCCI driver;
- AppDB: registering the virtual machine images that compose the platform.

The workflows have been successfully tested together with EGI Operations and the resource centres providing resources.

Furthermore, Terradue and EGI have agreed to two VO Service Level Agreements (EGI 2016) for the provisioning of EGI resource to support both the Geohazard and Hydrology TEPs. Thanks to such agreements, the GEP will be able to access European cloud compute resources. In total, the six data centres offer more than **360 virtual CPU cores**, **800 GB of memory** and **10 TB of storage**. The GEP just started to use the EGI Federated Cloud resource in production.

Conclusions

Nowadays, the implementation of the European Research Area (ERA), as depicted by the European Council, cannot be considered fully achieved. The realization of an open and integrated environment for cross-border seamless access to advanced digital resources, services and capabilities fostering the reuse of research data and services, is being accelerated by the European Open Science Cloud initiative of the European Commission. Open Science is seen as the natural paradigm to foster and drive such developments. It can remove the barrier between adjacent communities, enable multidisciplinary collaborations, reinforce the need of knowledge sharing and allow non-discriminatory access.

This chapter illustrated the benefits of a Commons approach to Open Science, and in particular the advantages of the Commons for the implementation of the European Open Science Cloud infrastructure and governance.

We presented a possible approach to implement the EOSC via the Open Science Commons. In our approach, the EOSC architecture will rely on a federation of cloud hubs, where a cloud hub provides data, services and capabilities in a standard and interoperable manner. The hubs adopt the cloud service provisioning paradigm to facilitate sharing, reusing and the combined offer of data and tools through the virtualisation. In addition, the federation of hubs provides a multilayer organizational structure that meets European policies, regulations, restrictions and business models and allows the creation of a community-network able to bring together the different kinds of expertise available in each hub. The multi-supply environment of the EOSC can be regulated by a corpus of policies, processes and tools defining the EOSC service integration and management system owned, maintained and developed by the EOSC according to the Commons governance model. The EOSC cloud hub services will be contributed by multiple stakeholders: data providers, European Research Infrastructures, e-Infrastructures, research collaborations and local, regional and national facilities.

The exploitation of Earth Observation data will directly benefit from the EOSC and the adoption of the Open Science Commons, leveraging technologies, services and resources provided in the context of existing European e-Infrastructures. EOSC and e-Infrastructures can become an attraction pole to design and develop appropriate solutions for managing and exploiting large amount of EO datasets. This will allow the creation of an integrated environment for fast development, prototyping and provisioning of services for exploitation platforms and scientific applications.

References

CYFRONET. One Data web site. https://www.onedata.org/
EGI (2016) VO SLA Terradue and VO OLAs Terradue. https://documents.egi.eu/document/2763
EGI (2017a). https://www.egi.eu/
EGI (2017b) EGI Service catalogue. https://www.egi.eu/services/
EGI (2017c) EGI Marketplace. http://marketplace.egi.eu/
EGI (2017d) EGI AppDB. http://appdb.egi.eu/
EGI (2017e) EGI Federated Cloud. https://www.egi.eu/federation/egi-federated-cloud/
EGI-Engage project (2017). http://cordis.europa.eu/project/rcn/194937_en.html
ESA (2016) ESA Thematic Exploitation Platforms architecture. http://go.egi.eu/EP-OpenArchitecture
EUDAT (2017). https://eudat.eu/
European Commission (2012) Communication from the Commission to the European Parliament, the Council, the European Economic and Social Committee and the Committee of the Regions, A Reinforced European Research Area Partnership for Excellence and Growth, COM (2012) 392 final. http://ec.europa.eu/euraxess/pdf/research_policies/era-communication_en.pdf. Accessed 15 Feb 2017
European Commission (2014a) Communication from the Commission to the Council and the European Parliament, European Research Area, Progress Report 2014. COM (2014) 575 final. http://ec.europa.eu/research/era/pdf/era_progress_report2014/era_progress_report_2014_communication.pdf. Accessed 15 Feb 2017
European Commission (2014b) Conclusions on progress in the European Research Area, Competitiveness Council meeting, Brussels, Feb 2014. http://www.consilium.europa.eu/uedocs/cms_data/docs/pressdata/en/intm/141120.pdf
European Commission (2015) Open Science at the Competitiveness Council. http://ec.europa.eu/digital-agenda/en/news/open-science-competitiveness-council-28-29-may-2015
European Commission (2016) First report of High Level Expert Group on the EOSC. https://ec.europa.eu/digital-single-market/en/news/first-report-high-level-expert-group-european-open-science-cloud
European Commission (2017) European Charter for Access to Research Infrastructures. https://ec.europa.eu/research/infrastructures/index_en.cfm?pg=access
European Commission. Science 2.0: science in transition. http://ec.europa.eu/research/consultations/science-2.0/background.pdf
European Strategy Forum on Research Infrastructures (2017) ESFRI web site. http://www.esfri.eu/. Accessed 15 Feb 2017

Benedikt   Fecher,   Sascha   Friesike   (2014)   Open   science: one   term,   five
    schools   of   thought.   http://book.openingscience.org/basics_background/
    open_science_one_term_five_schools_of_thought.html
Frischmann BM (2013) Infrastructure: the social value of shared resources. Oxford, Oxford
    University Press
GEANT (2017). http://www.geant.org/
Geohazard Exploitation Platform (2017). https://geohazards-tep.eo.esa.int/
Horizon 2020 consultation report (n.d.) Open Infrastructures for Open Science. http://
    cordis.europa.eu/fp7/ict/   e-Infrastructure/docs/open-infrastructure-for-open-science.pdf.
    Accessed 15 Feb 2017
e-Infrastructure Reflection Group (2012) e-IRG roadmap 2012. http://e-irg.eu/documents/10920/
    12353/e-irg_roadmap_2012-final.pdf/4c5cab85-dca4-49b7-b7f0-66c2bdd057af
e-Infrastructure Reflection Group (2017a) e-Infrastructure Commons. http://knowledgebase.e-
    irg.eu/e-infrastructure-commons
e-Infrastructure Reflection Group (2017b). http://e-irg.eu/
LIBER (2017). http://libereurope.eu/
OpenAIRE (2017). https://www.openaire.eu/
Mark D. Wilkinson et al. (2016) The FAIR Guiding Principles for scientific data management and
    stewardship. http://www.nature.com/articles/sdata201618

9

Fostering Cross-Disciplinary Earth Science through Datacube Analytics

Peter Baumann, Angelo Pio Rossi, Brennan Bell, Oliver Clements, Ben Evans, Heike Hoenig, Patrick Hogan, George Kakaletris, Panagiota Koltsida, Simone Mantovani, Ramiro Marco Figuera, Vlad Merticariu, Dimitar Misev, Huu Bang Pham, Stephan Siemen, and Julia Wagemann

With the unprecedented increase of orbital sensor, in situ measurement, and simulation data there is a rich, yet not leveraged potential for obtaining insights from dissecting datasets and rejoining them with other datasets. Obviously, goal is to allow users to "ask any question, any time, on any size", thereby enabling them to "build their own product on the go".

One of the most influential initiatives in EO is EarthServer which has demonstrated new directions for flexible, scalable EO services based on innovative NoSQL

P. Baumann • V. Merticariu • D. Misev
rasdaman GmbH, Bremen, Germany

Jacobs University, Bremen, Germany
e-mail: p.baumann@jacobs-university.de; v.merticariu@jacobs-university.de; d.misev@jacobs-university.de

A.P. Rossi • B. Bell • R. Marco Figuera • H.B. Pham
Jacobs University, Bremen, Germany
e-mail: an.rossi@jacobs-university.de; b.bell@jacobs-university.de; r.marcofiguera@jacobs-university.de; b.phamhuu@jacobs-university.de

O. Clements
Plymouth Marine Laboratory, Plymouth, UK
e-mail: olcl@pml.ac.uk

B. Evans
National Computational Infrastructure (NCI), Australian National University, Canberra, ACT, Australia
e-mail: Ben.Evans@anu.edu.au

H. Hoenig
rasdaman GmbH, Bremen, Germany
e-mail: hoenig@rasdaman.com

P. Hogan
NASA Ames, Moffett Field, CA, USA
e-mail: Patrick.Hogan@nasa.gov

technology. Researchers from Europe, the USA and Australia have teamed up to rigorously materialize the concept of the datacube. Such a datacube may have spatial and temporal dimensions (such as an *x/y/t* satellite image time series) and may unite an unlimited number of scenes. Independently from whatever efficient data structuring a server network may perform internally, users will always see just a few datacubes they can slice and dice.

EarthServer has established client and server technology for such spatio-temporal datacubes. The underlying scalable array engine, rasdaman, enables direct inter-action, including 3D visualization, what-if scenarios, common Earth Observation data processing, and general analytics. Services exclusively rely on the open OGC "Big Geo Data" standards suite, the Web Coverage Service (WCS). Phase 1 of EarthServer has advanced scalable array database technology into 100+ TB services; in Phase 2, Petabyte datacubes are being built for ad-hoc extraction, processing, and fusion.

But EarthServer has not only used, but also shaped several Big Data standards. This includes OGC coverage data and service standards, INSPIRE WCS, and the ISO Array SQL candidate standard.

We present the current state of EarthServer in terms of services and technology and outline its impact on the international standards landscape.

Introduction

The term "Big Data" is a contemporary shorthand characterizing data which are too large, fast-lived, heterogeneous, or complex to be understood and exploited. Technologically, this is a cross-cutting challenge affecting storage and processing, data and metadata, servers and clients and mash-ups. Further, making new, sub-stantially more powerful tools available for simple use by non-experts while not constraining complex tasks for experts just adds to the complexity. All this holds for many application domains, but specifically so for the field of Earth Observation (EO). With the unprecedented increase of orbital sensor, in situ measurement, and simulation data there is a rich, yet not leveraged potential for acquiring insights from dissecting datasets and rejoining them with other datasets. The stated goal is

G. Kakaletris • P. Koltsida
CITE s.a, Attiki, Greece
e-mail: g.kakaletris@cite.gr; p.koltsida@di.uoa.gr

S. Mantovani
MEEO s.r.l., Ferrara, Italy
e-mail: mantovani@meeo.it

S. Siemen • J. Wagemann
ECMWF, Reading, UK
e-mail: Stephan.Siemen@ecmwf.int; julia.wagemann@ecmwf.int

to enable users to "ask any question, any time, on any volume" thereby enabling them to "build their own product on the go".

In the field of EO, one of the most influential initiatives towards this goal is EarthServer (Baumann et al. 2015a; EarthServer 2015) which has demonstrated new directions for flexible, scalable EO services based on innovative NoSQL technology. Researchers from Europe, the USA and Australia have teamed up to rigorously materialize the concept of the datacube. Such a datacube can have spatial and temporal dimensions (such as a satellite image timeseries) and is able to unite an unlimited number of single images. Independent from whatever data structuring a server network may perform internally for efficiency on the millions of hyperspectral images and hundreds of climate simulations, users will always see just a few datacubes they can slice and dice.

EarthServer has established a slate of services for such spatio-temporal datacubes based on the scalable array engine, rasdaman, which enables direct interaction, including 3D visualization, what-if scenarios, common EO data processing, and general analytics. All services strictly rely on the open OGC data and service standards for "Big Geo Data", the Web Coverage Service (WCS) suite. In particular, the Web Coverage Processing Service (WCPS) geo raster query language has proven instrumental as a client data programming language which can be hidden behind appealing visual interfaces.

EarthServer has advanced these standards based on experience gained. The OGC WCS standards suite in its current, comprehensive state has been largely shaped by EarthServer which provides the Coverages, WCS, and WCPS standards editor and working group chair. The feasibility evidence provided by EarthServer has contributed to the uptake of WCS by open-source and commercial implementers; meanwhile, OGC WCS has been adopted by INSPIRE and has entered the adoption process of ISO.

Phase 1 of EarthServer has ended in 2014 (Baumann et al. 2015a); independent experts characterized the outcome, based on "proven evidence", that rasdaman will "significantly transform the way that scientists in different areas of Earth Science will be able to access and use data in a way that hitherto was not possible". And "with no doubt" this work "has been shaping the Big Earth Data landscape through the standardization activities within OGC, ISO and beyond". In Phase 2, which started in May 2015, this is being advanced even further: from the 100 TB database-size achieved in Phase 1 over the currently more than 500 TB, the next frontier will be crossed by building Petabyte datacubes for ad-hoc querying and fusion (Fig. 1).

In this contribution we present status and intermediate results of EarthServer and outline its impact on the international standards landscape. Further, we highlight opportunities established through technological advance and how future services can cope better with the Big Data challenge in EO.

The remainder of this contribution is organized as follows. In section "Standards-Based Modelling of Datacubes", the concepts of the OGC datacube and its service standards are introduced. An initial set of services in the federation is presented in section "Science Data Services", followed by an introduction to the

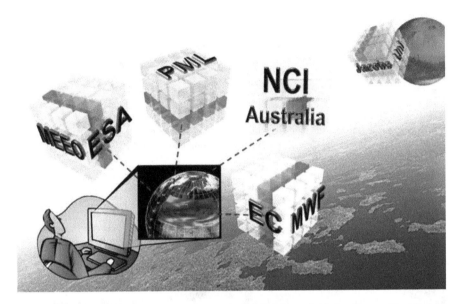

Fig. 1 Intercontinental datacube mix and match in the EarthServer initiative (Source: EarthServer)

underlying technology platform and an evaluation in section "Datacube Analytics Technology". Section ""Conclusion and Outlook concludes the plot with an outlook.

Standards-Based Modelling of Datacubes

EarthServer relies on the OGC "Big Earth Data" standards, WCS and WCPS, as the only client/server protocols for any kind of access and processing; additionally, WMS is offered. In the server, all such requests uniformly get mapped to an array query language which we will introduce later. Advanced visual clients enable point-and-click interfaces effectively hiding the query language syntax, except when experts want to make use of it. Additionally, access through expert tools like python notebooks is provided.

At the heart of the EarthServer conceptual model is the concept of *coverages* as digital representations of space/time varying phenomena as per ISO 19123 (ISO 2004) (which is identical to OGC Abstract Topic 6). Practically speaking, coverages encompass regular and irregular grids, point clouds, and general meshes. The datacube concept, being based on multidimensional arrays, represents a subset of coverages that focuses on regular and irregular spatio-temporal grids (Fig. 2).

The notion of *coverages* (Baumann 2012; Baumann and Hirschorn 2015; Wikipedia 2016) has proven instrumental in unifying spatio-temporal regular and

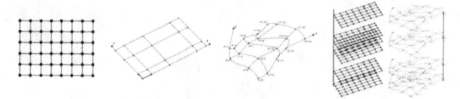

Fig. 2 Sample datacube grid types supported by rasdaman: regular grids (*left*), irregular and warped grids (*center left* and *right*), multidimensional combinations (*right*) of regular and irregular grid axes (Source: OGC/Jacobs University)

Fig. 3 WCS/WCPS based datacube services utilizing rasdaman (Source: rasdaman/EarthServer)

irregular grids, point clouds, and meshes so that such data can be accessed and processed through a simple, yet flexible and interoperable service paradigm.

By separating coverage data and service model, any service—such as WMS, WFS, SOS and WPS—can provide and consume coverages. That said, the *Web Coverage Service* (WCS) standard offers the most comprehensive, streamlined functionality (OGC 2016a). This modular suite of specifications starts with fundamental data access in WCS Core and has various extensions adding optionally implementable functionality facets, up to server-side analytics based on the *Web Coverage Processing Service* (WCPS) geo datacube language (Fig. 3) (Baumann 2010b).

Below we introduce the OGC coverage data and service model with an emphasis on practical aspects and illustrate how they enable high-performance, scalable implementations.

Coverage Data Model

According to the common geo data model used by OGC, ISO, and others, objects with a spatial (possibly temporal) reference are referred to as *features*. A special type of features are *coverages* whose associated values vary over space and/or time, such as an image where each coordinate leads to an individual color value. Complementing the (abstract) coverage model of ISO 19123 on which it is based, the (concrete) OGC coverage data and service model (Baumann 2012) establishes verifiable interoperability, down to pixel level, through the OGC conformance tests. While concrete, the coverage model still is independent from data format encodings—something which is of particular importance as it allows a uniform handling metadata, and individual mappings to the highly diverse metadata handling of the various data formats.

The OGC coverage model (and likewise WCS) meantime is supported by most of the respective tools, such as open-source MapServer, GeoServer, OPeNDAP and ESRI ArcGIS. In 2015, this successful coverage model has been extended to allow any kind of irregular grids, resulting in the OGC *Coverage Implementation Schema* (CIS) 1.1 (Baumann and Hirschorn 2015) which is in the final stage of adoption at the time of this writing. Different types of axes are made available for composing a multidimensional grid in a simple plug-and-play fashion. This effectively allows to concisely represent coverages ranging from unreferenced over regular grids to irregularly spaced axes (as often occurring in timeseries) and warped grids to ultimately algorithmically determined warpings, such as those defined by SensorML 2.0.

Web Coverage Service

The OGC service definition specifically built for deep functionality on coverages is the Web Coverage Service (WCS) suite of specifications. With WCS Core (Baumann 2010a), spatio-temporal subsetting as well as format encoding is provided; this Core must be supported by all implementations claiming conformance. Figure 4 illustrates WCS subsetting functionality, Fig. 5 shows the overall architecture of the WCS suite. Conformance testing of WCS implementations follows the same modularity approach and involves detailed checks, essentially down to the level of single cell (e.g. "pixel", "voxel") values (OGC 2016b). In December 2016, the European legal framework for a common Spatial Data Infrastructure, INSPIRE, has adopted WCS as coverage download service (INSPIRE 2016).

Fig. 4 WCS subsetting: trimming (*left*) and slicing (*right*) (Source: OGC)

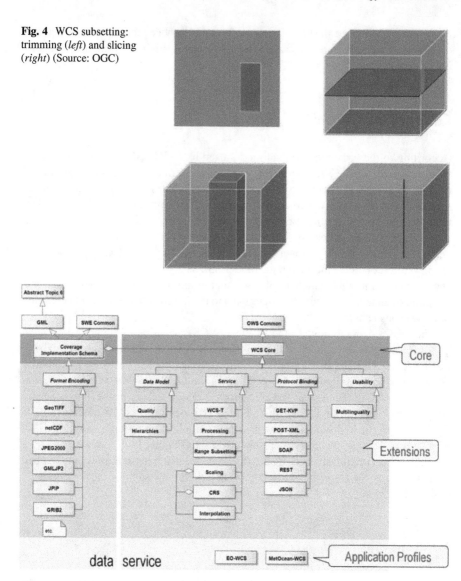

Fig. 5 Overall WCS suite architecture (Source: OGC)

Web Coverage Processing Service

Web Coverage Processing Service (WCPS) is OGC's geo raster query language, adopted already in 2008 (Baumann 2010b). An example may illustrate the use of WCPS: "*From MODIS scenes M1, M2, M3, the difference between red & near-infrared bands, encoded as TIFF—but only those where near-infrared exceeds threshold 127 somewhere.*" The corresponding query reads as follows:

Fig. 6 3D rendering of datacube query results (data and service: BGS, server: rasdaman)

```
for $c in doc("http://acme.com/wcs")//coverage
where  some( $c.nir > 127 )
return encode( $c.red - $c.nir, "image/tiff" )
```

Such results can conveniently be rendered through WebGL in a standard Web browser, or through NASA WorldWind (Fig. 6). The syntax is close to SQL/MDA (see below), but with a syntax flavor close to XQuery so as to allow integration with XPath and XQuery, for which a specification draft is being prepared by EarthServer (see section on data/metadata integration further down).

The Role of Standards

As the hype dust settles down over "Big Data" the core contributing data structures and their particularities crystallize. In Earth Science data, these arguably are regular and irregular grids, point clouds, and meshes, reflected by the coverage concept. The unifying notion of coverages appears useful as an abstraction that is independent from data formats and their particularities while still capturing the essentials of spatio-temporal data. With CIS 1.1, description of irregular grids has been simplified by not looking at the grids, but at the axis characteristics. While many services on principle can receive or deliver coverages, the WCS suite is specifically designed to not only work on the level of whole (potentially large) objects, but can address inside objects as well as filter and process them, ranging up to complex analytics with WCPS.

The critical role of flexible, scalable coverage services for spatio-temporal infrastructures is recognized far beyond OGC, as the substantial tool support highlights. This has prompted ISO and INSPIRE to also adopt the OGC coverage and WCS standards. Also, ISO is extending the SQL standard with n-D arrays (ISO 2015; Misev and Baumann 2015). The standards observing group of the US Federal Geographic Data Committee (FGDC) sees coverage processing a la WCS/WCPS as a future "mandatory standard". In parallel, work is continuing in OGC towards extending coverage world with further data format mappings and to add further relevant functionality, such as flexible XPath-based coverage metadata retrieval. Finally, research is being undertaken on embedding coverages into the Geo Semantic Web (Baumann et al. 2015b), also supporting W3C which has started studying coverages in the "Spatial Data on the Web" Working Group. A demonstration service for 1D through 5D coverages is available for studying the WCS/WCPS universe (rasdaman 2016a).

Science Data Services

Expertise of the EarthServer project partners covers multiple scientific domains. This ensures that benefits achieved can be made available to the largest possible audience. Partners include Plymouth Marine Laboratory (PML) running the Marine Data Service, Meteorological and Environmental Earth Observation (MEEO) operating the Sentinel Earth Observation Service, National Computing Infrastructure (NCI) Australia running the LandSat Earth Observation Service, European Centre for Medium-Range Weather Forecasts (ECMWF) with its Climate Data Service, and Jacobs University providing the Planetary Data Service. Based on the common EarthServer platform provided by the technology partners Jacobs University, rasdaman GmbH, CITE s.a., and NASA all the aforementioned service partners have set up domain specific clients and data access portals which are continuously advanced and populated over the lifetime of the project so as to cross the Petabyte frontier for single services in 2017. Multiple service synergies will be explored which will allow users to query and analyze data stored at different project partner's infrastructure from a single entry point. An example of this is the LandSat service being developed jointly by MEEO and NCI. The specific data portals and access options are detailed in the following sections.

Earth Observation Data Services

The use of EO data is getting more and more challenging with the advent of the Sentinel era. The free, full and open data policy adopted for the Copernicus programme foresees access available to all users for the Sentinel data products. Terabytes of data and EO products are already generated every day from Sentinel-1 and Sentinel-2, and with the approaching launch of the Sentinel-3/-4/-5P/-5, the need of advanced access services is crucial to support the increasing data demand from the users.

The Earth Observation Data Service (Alfieri et al. 2013; MEEO 2017) offers dynamic and interactive access functionalities to improve and facilitate the accessibility to massive Earth Science data: key technologies for data exploitation (Multi-sensor Evolution Analysis (2016), (rasdaman 2016b), NASA Web World Wind (NASA 2016)) are used to implement effective geospatial data analysis tools empowered with the OGC standard interfaces for Web Map Service (WMS) (De la Beaujardiere 2016), Web Coverage Service (WCS) (Baumann 2010a), and Web Coverage Processing Service (WCPS) (Baumann 2008)—see Fig. 7.

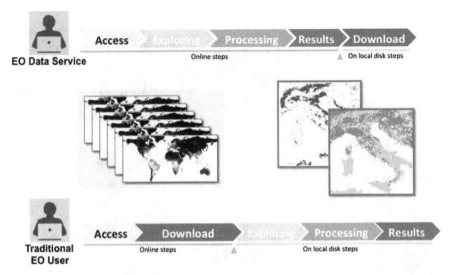

Fig. 7 Data exploitation approaches offered by traditional (*bottom*) and EO Data Service (*top*) approaches (Source: MEEO)

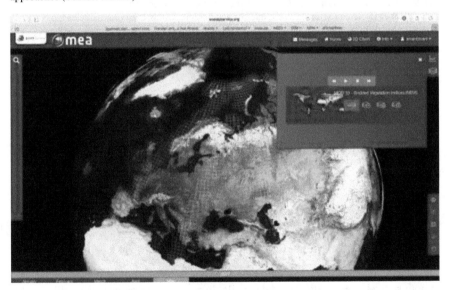

Fig. 8 Availability of global Sentinel 2A data on May 19th, 2017 on top of MODIS normalize difference vegetation index (NDVI) visualized in the ESA/NASA Web World Wind virtual globe (Source: MEEO)

With respect to the traditional data exploitation approaches, the EO Data Service supports on-line data interaction, restructuring the typical steps and moving to the end the download of the real data of interest for the users with a significant reduction of data transfer (Figs. 8 and 9).

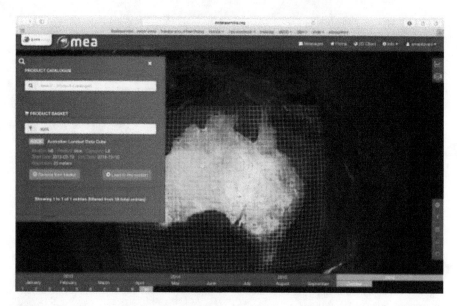

Fig. 9 Australian Landsat Data Cube coverage, presented in the ESA/NASA WebWorldWind virtual globe. Users can select areas of interest and explore Landsat data available at the National Computational Infrastructure (NCI) Australia (Source: MEEO)

The EO Data Service currently provides in excess of one PB of ESA and NASA EO products (e.g. vegetation indexes, land surface temperature, precipitation, soil moisture, etc.) to support Atmosphere, Land and Ocean applications.

In the framework of the EarthServer initiative, the Big Data Analytics tools are being enabled on datacubes of Copernicus Sentinel and Third Party Missions (e.g. Landsat8) data, coming from MEEO and its federation partner NCI Australia, to support agile analytics and fusion on these new generation sensors through MEEO's service (Fig. 10).

Marine Science Data Service

The marine data service (Marine Data Service) is focused on providing access to remote sensed ocean data. The data available are from ocean colour satellites. The marine research community is well accustomed to using satellite data. Satellite data provides many benefits over in situ observations. The data have a global coverage and provide a consistent and accurate time series of data. The marine research community has recognized the benefit of long time series of data. Time series need to be consistent so that the data are comparable through the whole series. Remote sensed data have helped to provide this consistency.

The ESA OC-CCI project (Sathyendranath et al. 2012) is producing a time series of remote sensed ocean colour parameters and associated uncertainty variables.

Fig. 10 Screenshot showing GIS client displaying chlorophyll data selected based on the per pixel value of uncertainty criteria, together with corresponding WCPS query (*left*) (Source: PML)

Currently the available time series runs from 1997 to 2015 and represents 1 of 14 subgroups of the overall ESA CCI project. With the creation of these large time series an increasingly technical challenge has emerged, how do users get benefit from these huge data volumes?

The EarthServer project, through the use of a suite of technologies including rasdaman and several OGC standard interfaces, aims to address the issue of users having to transfer and store large data volumes by offering ad-hoc querying over the whole data catalog.

Traditionally a marine researcher would simply select the particular temporal and spatial subset of the dataset they require from a web based catalog and download to their local disk. This system has worked well but is becoming less feasible due to the increases in data volume and the increase in non-specialists wanting access to the data. Take for example a researcher interested in finding the average monthly chlorophyll concentration for the North Sea for the period 2000–2010. Traditional methodologies would require the download of around several gigabytes of data. This represents a large time investment for the actual download as well as a cost associated with storage and processing required (Clements and Walker 2014). By making the same dataset available through the EarthServer project a research can simply write the analysis as a WCPS query and send that to the data service. The analysis is done at the data and only the result is downloaded, in this case around 100 KB. This example outline the clearest cut advantage, however there are more transient benefits that could improve the way that researchers interact with and use data. One example of this would be the testing of novel algorithms that require access to the raw light reflectance data. These data are used through existing algorithms to calculate derived products such as chlorophyll concentration, primary production and carbon sequestration.

The marine data service currently provides in excess of 70 TB of data. Through the course of the project we will be expanding the data offering to include data from the ESA Sentinel 3 Ocean and Land Colour Instrument (OLCI) (Berger et al. 2012). The aim is to offer as much data from the sensor as is available with the total goal to offer 1 PB of data through the service.

Climate Science Data Service

The Climate Science Data Service is developed by the European Centre for Medium-Range Weather Forecasts (ECMWF). ECMWF hosts the Meteorological Archival and Retrieval System (MARS), the largest archive of meteorological data worldwide with currently more than 170 PB of data (ECMWF 2014). As a Numerical Weather Predication (NWP) centre, ECMWF primarily supports the meteorological community through well-established services for accessing, retrieving and processing data from the MARS archive. User outside the MetOcean domain, however, often struggle with the climate-specific conventions and formats, e.g. the GRIB data format. This limits the overall uptake of ECMWF data. At the same time, with data volumes in the range of Petabytes, data download for processing on users' local workstations is no longer feasible. ECMWF as a data provider has to find solutions to provide efficient web-based access to the full range of data while at the same time the overall data transport is minimized. Ideally, data access and processing takes place on the server and the user only downloads the data that is really needed.

ECMWF's participation in EarthServer-2 aims at addressing exactly this challenge: to give users access to over 1 PB of meteorological and hydrological data and at the same time providing tools for on-demand data analysis and retrieval. The approach is to connect the rasdaman server technology with ECMWF's MARS archive, thereby enabling access to global reanalysis data via the OGC-based standards Web Coverage Service (WCS) and Web Coverage Processing Service (WCPS). This way, multidimensional gridded meteorological data can be extracted and processed in an interoperable way.

The climate reanalysis service particularly addresses users outside the MetOcean domain, more familiar with common Web and GIS standards. A WC(P)S for climate science data can be of benefit for developers or scientists building Web-applications based on large data volumes, who are unable to store all the data locally. Technical data users for example can integrate a WCS request into their processing routine and further process the data. Companies can easily build customized web-applications with data provided via a WCS. This approach is also strongly promoted by the EU's Copernicus EO programme which generates climate and environmental data as part of its operational services. Companies can use these data for value-added climate services for decision-makers or clients (Fig. 11).

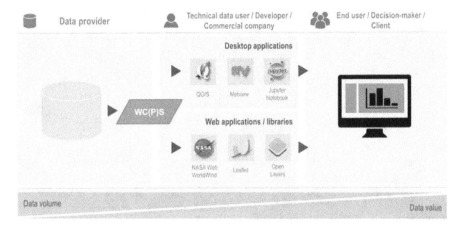

Fig. 11 Example of how a WC(P)S can be integrated into standard processing chains (Source: ECMWF)

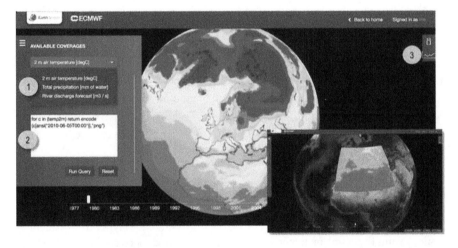

Fig. 12 WebWorldWind client, with three main functionalities: (*1*) 3D visualization, (*2*) writing own WCPS queries to choose a coverage subset (compare inlet) and (*3*) plotting of time series/hydropgraph of selected latitude/longitude information (Source: ECMWF)

To showcase how simple it is to build a custom web application with the help of a WC(P)S, a demo web client visualizing ECMWF data with NASA WorldWind has been developed (ECMWF n.d.) giving access to currently three datasets: ERA-interim 2 m air temperature and total accumulated precipitation (Dee et al. 2011) as well as GloFAS river discharge forecast data (Alfieri et al. 2013). Two-dimensional global datasets can be mapped on the globe (Fig. 12). An additional plotting functionality allows retrieval of data points in time for individual coordinates. This is suitable for ERA-interim time-series data and hydrographs based on river-discharge forecast data (Fig. 13).

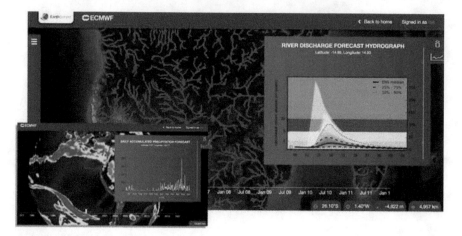

Fig. 13 Sample plotting functionalities. The main image shows a hydrograph plotted based on daily river discharge forecast data. The inlet shows plotting of ERA-interim time series data. The plot shows the total accumulated precipitation for one lat/lon grid point for 1 year (Source: Jacobs University)

In summary, WCS for Climate Data offers a facilitated on-demand access to ECMWF's climate reanalysis data for researchers, technical data users and commercial companies, within the MetOcean community and beyond.

Planetary Science Data Service

Planetary Science missions are largely based on Remote Sensing experiments, whose data are very much comparable with those from Earth Observation sensors. Data are thus relatively similar in terms of data structure and type: from panchromatic, to multispectral or hyperspectral data, as well as derived datasets such as stereo-derived topography, or laser altimetry, in terms of surface imaging (Oosthoek et al. 2013), in addition to subsurface vertical radar sounding (Cantini et al. 2014), or atmospheric imaging and profiles. The vast majority of these data can be represented with raster models, thus they are suitable for use in array databases.

Planetary raster data have never much suffered from being closed in archives during last decades: all remote sensing imagery returned by spacecrafts is available in the public domain, together with documentation (e.g. (Heather et al. 2013; McMahon 1996)). Nevertheless, archived data are typically lower-level, unprocessed or partially processed images and cubes, not GIS- and science-ready products. In addition, they typically are analyzed as single data granules or with cumbersome processing and analyzing pipelines to be carried out by individual scientists, on own infrastructure.

What is also slightly challenging for the access, integration and analysis of Planetary Science data is the wide range of bodies in terms of surface (or atmosphere) nature, experimental characteristics and Coordinate Reference Systems. The sheer volume of data, counted in few GB for entire missions (such as NASA Viking orbiters) until the 1980s, is now approaching the order of magnitude of tens to hundreds of TB.

All these aspects tend to give a Big Data dignity to Planetary datasets, too. The planetary community expressed the need during the past decade of easier and more effective ways to access and analyze its wealth data (Pondrelli et al. 2011). Most web services to date addressed the availability of maps (e.g. with WMS), but not extensively the deeper access, in terms of analysis and analytics to the complexity and richness of planetary datasets. WCPS demonstrated the capability to address this (Oosthoek et al. 2013; Rossi et al. 2014).

The Planetary Science Data Service (PSDS) of EarthServer, also known as PlanetServer (2016a), focuses on complex multidimensional data, in particular hyperspectral imaging and topographic cubes and imagery. All of those data derive from public archives and are processed to the highest level with publicly available routines.

In addition to Mars data (Rossi et al. 2014), WCPS is offered on diverse datasets on the Moon, as well as Mercury. Other Solar System Bodies are also going to be covered and served. Derived parameters such as hyperspectral summary products and indices can be produced through WCPS queries, as well as derived imagery color combination products.

One of the objectives of PlanetServer is to translate scientific questions into standard queries that can be posed to either a single granule/coverage, or an extremely large number of them, from local to global scale. The planetary and remote sensing and geodata communities at large could benefit from PlanetServer at different levels: from accessing its data and performing analyses with its web services, for research or education purposes; to using and adapting or iterating further the concepts and tools developed within PlanetServer.

PlanetServer in its new iteration is completely based on open source software, and its code available on GitHub (PlanetServer 2016b). The main server component empowering PlanetServer-2 is rasdaman community edition, and its visualization engine is the NASA WorldWind virtual globe (Fig. 14) (Hogan 2011).

A sample, nontrivial WCPS query for returning an RGB combination from MRO CRISM hyperspectral imaging data for the mineral compositional indices *sindex2*, *BD2100_2*, and *BD1900_2* mapped to RGB as described by Viviano-Beck et al. (2014) is the following one, with null values set to transparent:

```
for data in (last_ingestion_2)
return encode(
  {
    red:    (int)(255/(max(data.band_233)-min(data.band_233)))
              * (data.band_233 - min(data.band_233));
    green:  (int)(255/(max(data.band_13)-min(data.band_13)))
              * (data.band_13 - min(data.band_13));
    blue:   (int)(255/(max(data.band_78)-min(data.band_78)))
```

Fig. 14 PlanetServer showing a Mars globe based on Viking Orbiter imagery mosaics produced by the United States Geological Survey (USGS), served from its rasdaman database draped on the WebWorldWind virtual globe using mosaicked NASA LRO mission data (Source: Jacobs University).

```
                    * (data.band_78 - min(data.band_78)) ;
    alpha: (int)(data.band_100 > 0) * 255
},
"png", "nodata=65535"
)
```

The result of this query is a map-projected subset of a cube highlighting compositional variations on the Surface (Fig. 15).

Cross-Service Federation Queries

Among the features of the EarthServer platform, consisting of metadata-enhanced rasdaman (see next subsection), is the capability to federate services. Technically, this is only a generalization of the service internal parallelization and distributed processing; externally, it achieves location transparency allowing users to send any query to any data center, regardless of which data are accessed and possibly combined, including across data center boundaries.

A lab prototype of this federation has been demonstrated live at EGU 2015 and AGU 2016 where a nontrivial query required combination of climate data from ECMWF in the UK with LandSat 8 imagery at NCI Australia. This query was alternately sent to ECMWF and NCI; each of the receiving services forked a

Fig. 15 WCPS query result
from the RGB combination
red: sindex2, green:
BD2100_2, blue: BD1900_2
(Source: PlanetServer)

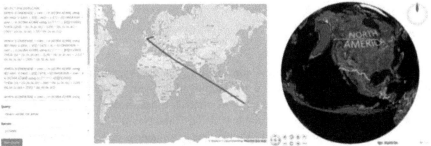

Fig. 16 Visualization of query splitting: original query (*left*), query distribution from Germany to the UK, with subquery spawned to Australia (*center*), query result visualized in NASA WorldWind (Source: EarthServer)

subquery to the service holding the data missing locally. The result was displayed in NASA WebWorldWind, allowing to visually assess equality of the results. Figure 16 shows part of the query, a visualization of the path the query fragments take, and the final result mapped to a virtual globe.

Datacube Analytics Technology

EarthServer uses a combination of Big Data storage, processing, and visualization technologies. In the backend, this is the rasdaman Array Database system which we introduce in the next section. Data/metadata integration plays a crucial role in the

EarthServer data management approach and is presented next. Finally, the central visualization tool, the NASA WorldWind virtual globe, is presented.

Array Databases as Datacube Platform

The common engine underlying EarthServer is the rasdaman Array Database (Baumann et al. 1999). It extends SQL with support for massive multidimensional arrays, together with declarative array operators which are heavily optimized and parallelized (Dumitru et al. 2014) on server side. A separate layer adds geo semantics, such as knowledge about regular and irregular grids and coordinates, by implementing the OGC Web service interfaces. For OGC and INSPIRE WCS, as well as OGC WCPS, rasdaman acts as reference implementation. On storage, arrays get partitioned ("tiled") into sub-arrays which can be stored in a database or directly in files. Additionally, rasdaman can access preexisting archives by only registering files, without copying them. Figure 1 shows the overall architecture of rasdaman.

Array Storage

Arrays are maintained in either a conventional database (such as PostgreSQL) or its own persistent store directly in any kind of file system. Additionally, rasdaman can tap into "external" files not under its control. Since rasdaman 9.3, an internal tiling of archive files (such as available with TIFF and NetCDF, for example) can be exploited for fine-grain reading. Under work is automated distribution of tiles based on various criteria, optionally including redundancy (Fig. 17).

A core concept of array storage in rasdaman is partitioning or *tiling*. Arrays are split into sub-arrays called *tiles* to achieve fast access. Tiling policy is a tuning parameter which allows adjusting partitions to any given query workload, measured or anticipated. As this mechanism turned out very powerful for users, its generality has been cast into a few strategies available to data designers (Fig. 18).

Array Processing

The rasdaman server ("rasserver") is the central workhorse. It can access data from various sources for multi-parallel, distributed processing. The rasdaman engine has been crafted from scratch, optimizing every single component for array processing. A series of highly effective optimizations is applied to queries, including:

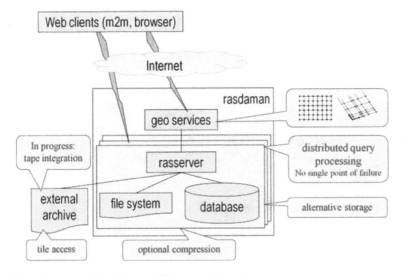

Fig. 17 rasdaman overall architecture (Source: rasdaman)

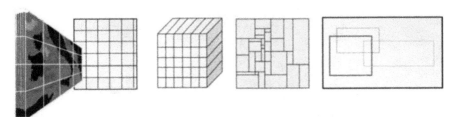

Fig. 18 Sample tiling rasdaman strategies supported (Source: rasdaman)

- Query rewriting to find more efficient expressions of the same query; currently 150 rewriting rules are implemented.
- Query result caching is used to keep complete or partial query results in (shared) memory for reuse by subsequent queries; in particular, geographic or temporal overlap can be exploited.
- Array joins with optimized tile loading so as to minimize multiple loads when combining two arrays (Baumann and Merticariu 2015). This is not only effective in a local situation, but also when tiles have to be transported between compute nodes or even data centers in case of a distributed join.

After query analysis and optimization, the system fetches only the tiles required for answering the given query. Subsequent processing is highly parallelized. Locally, it assigns tiles to different CPUs and threads. In a cluster, query are split and parallelized across the nodes. The same mechanism is also used for distributing processing across data centers, where data transport becomes a particular issue. To maximize efficiency, rasdaman currently optimizes splitting along two criteria

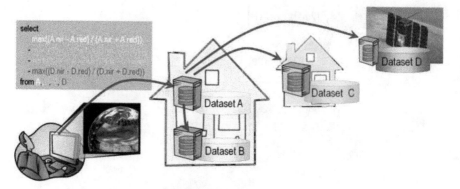

Fig. 19 rasdaman query splitting (Source: rasdaman)

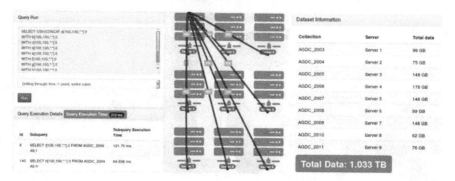

Fig. 20 Visualization workbench for rasdaman distributed query processing (Source: rasdaman)

(Fig. 19): First, send queries to where the data sit ("shipping code to data"); second, generate subqueries that process as much as ever possible locally, minimizing the amount of data to be transported between nodes.

This way, single queries have been successfully split across more than a thousand Amazon cloud nodes (Dumitru et al. 2014). Figure 20 shows an experiment done on the rasdaman distributed query processing visualization workbench where nine Amazon nodes process a query on 1 TB processed in 212 ms.

Tool Integration

Even though the WCS, WCS, and WCPS protocols are open, adopted standards, they are not necessarily appropriate for end users—from WMS we are used to have Web clients like OpenLayers and Leaflet which hide the request syntax, and the same holds for WCS requests and, although high-level and abstract, the WCPS language. In the end, all these interfaces are most useful as client/server communication protocols where end users are hidden from the syntax through visual point-and-click interfaces (like OpenLayers and NASA WorldWind) or, alternatively, through their own, well-known tools (like QGIS and python).

To this end, rasdaman already supports major GIS Web and programmatic clients, and more are under development. Among this list are MapServer, GDAL, EOxServer, OpenLayers, Leaflet, QGIS, and NASA WorldWind, C++, and Java. Python is in advanced development stage.

The Role and Handling of Metadata

Metadata can be of utmost importance for the utilization of datasets, as apart from textual descriptions and provenance traces, it may provide essential information on how data may be consumed or interpreted (e.g. characteristics of equipment/process, reference systems, error margins). When data management crosses the boundaries of systems, institutions and scientific disciplines, metadata management becomes a complex process on its own. The Earth-Sciences landscape is an ample example where datasets, which are substantially "many", may be considered from a variety of standpoints, and be produced/consumed by heterogeneous processes in various disciplines with diverse needs and concepts.

Focusing on coverages hosted behind WCS and WCPS services, where metadata heterogeneity is evident due to the liberal approach of the relevant specifications, the EarthServer 2 metadata management system addresses the challenge, by being metadata schema agnostic yet maintaining the ability to host and process composite metadata models. Meanwhile, the system seeks to meet a number of supplementary requirements such as fault-tolerance, efficiency and scalability, looking to a (near) future where hosting billions of datasets will be common case.

The system supports of two modes of operation, with quite distinct characteristics (a) *in situ operation* (metadata are not relocated and services are offered on top of the original store's metadata retrieval ones) and (b) *federated operation* (metadata are gathered in a distributed store over which the full range of system services may be provided).

The architecture (cf. Fig. 21) consists of loosely coupled distributed services that interoperate through standards, WCS and WCPS being the fundamental ones. XPath is utilized for metadata retrieval/filtering, over NoSQL technologies in order to achieve the desired scalability, performance and functional characteristics. Full text queries are also supported. In federated mode, services are invoked using WCPS or WCS-T standards. Other supported protocols include OpenSearch, OAI-PMH and CSW.

Access to the combined processing and retrieval engine is provided via *xWCPS2.0*, a specification that leverages the agile earth-data analytics layer with effective metadata retrieval and processing facilities, delivering an expressive

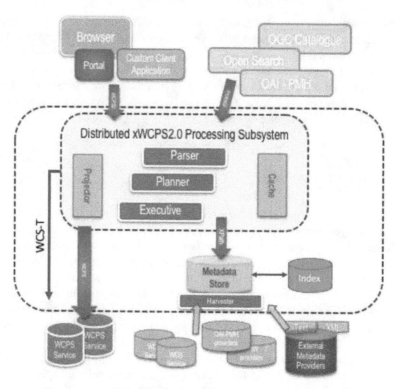

Fig. 21 xWCPS overall architecture (Source: CITE).

querying tool that can interweave data and metadata in composite operations. xWCPS 2.0 builds on xWCPS1.0 (from EarthServer-1) and, apart from an enhanced FLWOR syntax, it delivers features that significantly enhance the ability to issue federated queries.

In the following xWCPS2.0 example, coverages across all federated servers ("*") are located via their metadata (name of <field> is elevation in where clause) and results consist of xml elements (<result> in return clause), containing the outcome of an XPath expression (metadata) and a WCPS evaluated element (value):

```
for    $c in *
where $c::://*[local-name()='field'][@name=elevation]
return
 <result>
   <value>
         $c[Lat(53.08),Long(8.80),ansi("2014-01":"2014-12")]
   </value>
   <metadata>$c:://domainSet</metadata>
 </result>
```

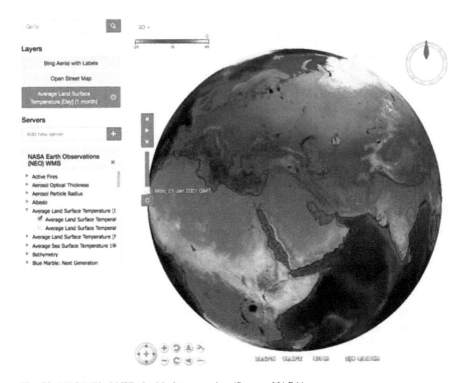

Fig. 22 NASA World Wind with data mapping (Source: NASA).

Virtual Globes as Datacube Interfaces

Visual globes help users experiencing their data visually with the various aspects displayed in their native context. This allows data to be more easily understood and their impacts better appreciated.

NASA is a pioneer in virtual globe technology, substantially preceding tools such as Google Earth. Our primary mission has always been to support the operational needs of the geospatial community through a versatile open source toolkit, versus a closed proprietary product. A particular feature of WorldWind is its modular and extensible architecture. WorldWind as an Application Programming Interface, API-centric Software Development Toolkit (SDK) can be plugged into any application that has spatial data needing to be experienced in the native context of a virtual globe (Fig. 22).

In EarthServer, the virtual globe paradigm is coupled with the flexible query mechanism of databases. Users can query rasdaman flexibly and have the results mapped to the globe. Rasdaman applications can add any 2D, 3D or 4D information to the WorldWind geobrowser for any dynamically generated query result. This enables a direct interaction with massive databases, as the excerpt of interest is prepared in the server while WorldWind accomplishes sophisticated interactive visualization in the native context of Earth as observed from space, thereby providing access to the various thematic EarthServer databases; with PlanetServer, WorldWind is also used for Mars, Moon and further solar system bodies.

Related Work

A large, growing number of both open-source and proprietary implementations is supporting coverages and WCS (Fig. 3). Specifically, the most recent version (OGC Coverage Implementation Schema 1.0 and WCS 2.0) are known to be implemented by open-source rasdaman (2016b), GDAL, QGIS, OpenLayers, Leaflet, OPeNDAP, MapServer, GeoServer, GMU, NASA WorldWind, EOxServer as well as proprietary Pyxis, ERDAS and ArcGIS. The most comprehensive tool is rasdaman—also OGC WCS Core Reference Implementation—which implements WCS Core and all extensions, including WCPS. This large adoption basis of OGC's coverage standards promotes interoperability of EarthServer with other services, supporting the GEOSS "system of systems" approach (Christian 2005). Notably, rasdaman is part of the GCI (GEOSS Common Infrastructure) (GEOSS 2016).

Google Earth Engine (Google n.d.) builds on the tradition of Grid systems. Users can submit python code which is executed transparently in a distributed processing environment. However, procedural code does not parallelize easily, therefore—after discussion with the rasdaman team—developers have added a declarative "Map Algebra" interface in addition which resembles a subset of an array query language. Still, many common techniques (like query compilation, heuristic rewriting, cost-based optimizations, adaptive tiling, data compression, etc.) are not being utilized—in the end, a substantial advantage comes from using the massive underlying Google hardware.

SciDB is an Array Database prototype under development (Paradigm4 2016) with no specific geo data support like OGC WCS interfaces. SciQL is a concept study adding arrays to a column store (Zhang et al. 2011). A performance comparison between rasdaman, SciQL, and SciDB shows that rasdaman excels by one, often several orders of magnitude in performance and also conveys better storage efficiency (Merticariu et al. 2015). To the best of our knowledge, only rasdaman has publicly available services deployed (Baumann et al. 2015b). No particular SciDB support for Earth data is known—the only supported ingest format is CSV (comma-separated values), and geo semantics is not available in queries.

Sensor Observation Service (SOS) supports delivery of sensor data (Bröring et al. 2012) which can be imagery. However, there is rather limited functionality, and performance is reported as not entirely satisfactory.

OGC WMTS exposes tiling to clients for maximizing performance (Masó et al. 2010); on the downside, queries are fixed to retrieval of such tiles, i.e. there is no free subsetting and no processing. OGC WPS provides an API for arbitrary processing functionality, however, is not interoperable per se as stated already in the standard (Schut 2007).

In ISO, an extension to SQL is in advanced stage which adds n-D arrays in a domain-independent manner (ISO 2015). SQL/MDA (for "Multidimensional Arrays") has been initiated by the rasdaman team, which also has submitted the specification; see (Misev and Baumann 2015) for a condensed overview. Adoption is anticipated for summer 2017.

Conclusion and Outlook

Datacubes are a convenient model for presenting users with a simple, consolidated view on the massive amount of data files gathered—"a cube tells more than a million images". Such a datacube may have spatial and temporal dimensions (such as a satellite image time series) and may unite an unlimited number of individual images. Independently from whatever efficient data structuring a server network may perform internally, users will always see just a few datacubes they can slice and dice.

Following the broadening of minds through the NoSQL wave, database research has responded to the Big Data deluge with new data models and scalability concepts. In the field of gridded data, Array Databases provide a disruptive innovation for flexible, scalable data-centric services on datacubes. EarthServer exploits this by establishing a federation of services of 3D satellite image timeseries and 4D climatological data where each node can answer queries on the whole network, in a federation implementing a "datacube mix and match". While in Phase 1 of EarthServer the 100 TB barrier has been transcended, in its Phase 2 it is attacking the Petabyte frontier.

Aside from using the OGC "Big Geo Data" standards for its service interfaces, EarthServer keeps on shaping datacube standards in OGC, ISO, and INSPIRE. Current work involves implementation of the OGC coverage model version 1.1, supporting data centers in establishing rasdaman-based services, and enhancing further the data and processing parallelism capabilities of rasdaman.

Acknowledgement The EarthServer initiative is partly funded by the European Commission under grant agreement FP7 286310 and H2020 654367.

References

Alfieri L et al (2013) GloFAS - global ensemble streamflow forecasting and flood early warning. Hydrol Earth Syst Sci 17:1161–1175

Andrejev A, Baumann P, Misev D, Risch T (2015) Spatio-temporal gridded data processing on the semantic web. IEEE Intl. Conf. on Data Science and Data Intensive Systems (DSDIS 2015), Sydney, Australia, December 11–13, 2015

Baumann P (2008) OGC Web Coverage Processing Service (WCPS) Language, version 1.0. OGC document 08-068r2. www.opengeospatial.org/standards/wcps. Accessed 21 May 2017

Baumann P (2010a) OGC WCS interface standard – core, version 2.0. OGC document 09-110r4. www.opengeospatial.org/standards/wcs. Accessed 21 May 2017

Baumann P (2010b) The OGC web coverage processing service (WCPS) standard. Geoinformatica 14(4):447–479

Baumann P (2012) OGC Coverage Implementation Schema (formerly: GML 3.2.1 application schema – coverages), version 1.0. OGC document 09-146r2. www.opengeospatial.org/standards/wcs. Accessed 21 May 2017

Baumann P, Hirschorn E (2015) OGC coverage implementation schema, version 1.1, OGC document 09-146r3

Baumann P, Merticariu V (2015) On the efficient evaluation of array joins. Proc. IEEE Big Data Conf. Workshop Big Data in the Geo Sciences, Santa Clara, US, October 29, 2015

Baumann P, Dehmel A, Furtado P, Ritsch R, Widmann N (1999) Spatio-temporal retrieval with RasDaMan. Proc. VLDB, Edinburgh, Scotland, UK, September 7–10, 1999

Baumann P, Mazzetti P, Ungar J, Barbera R, Barboni D, Beccati A, Bigagli L, Boldrini E, Bruno R, Calanducci A, Campalani P, Clement O, Dumitru A, Grant M, Herzig P, Kakaletris G, Laxton J, Koltsida P, Lipskoch K, Mahdiraji A, Mantovani S, Merticariu V, Messina A, Misev D, Natali S, Nativi S, Oosthoek J, Passmore J, Pappalardo M, Rossi A, Rundo F, Sen M, Sorbera V, Sullivan D, Torrisi M, Trovato L, Veratelli M, Wagner S (2015a) Big data analytics for earth sciences: the EarthServer approach. Int J Dig Earth :1–27

Baumann P, Mazzetti P, Ungar J, Barbera R, Barboni D, Beccati A, Bigagli L, Boldrini E, Bruno R, Calanducci A, Campalani P, Clement O, Dumitru A, Grant M, Herzig P, Kakaletris G, Laxton J, Koltsida P, Lipskoch K, Mahdiraji AR, Mantovani S, Merticariu V, Messina A, Misev D, Natali S, Nativi S, Oosthoek J, Passmore J, Pappalardo M, Rossi AP, Rundo F, Sen M, Sorbera V, Sullivan D, Torrisi M, Trovato L, Veratelli MG, Wagner S (2015b) Big data analytics for earth sciences: the EarthServer approach. Int J Dig Earth :1–27

Berger M, Moreno J, Johannessen J, Levelt P (2012) Hanssen R (2012) ESA's sentinel missions in support of Earth system science. Remote Sens Environ 120:84–90

Bröring A, Stasch C, Echterhoff J (2012) OGC® sensor observation service interface standard, version 2.0. OGC document 12-006

Cantini F, Rossi A, Orosei R, Baumann P, Misev D, Oosthoek J, Beccati A, Campalani P, Unnithan V (2014) MARSIS data and simulation exploited using array databases: PlanetServer/EarthServer for sounding radars. Geophys Res Abstr 16:EGU2014-3784

Christian E (2005) Planning for the global earth observation system of systems (GEOSS). Space Policy 21(2):105–109

Clements O, Walker P (2014) Can EO afford big data – an assessment of the temporal and monetary costs of existing and emerging big data workflows. Proc. EGU, Vienna, Austria, April 27–May 02, 2014

De la Beaujardiere J (2016) Web Map Service (WMS), version 1.3. OGC document 06-042. www.opengeospatial.org/standards/wms. Accessed 21 May 2017

Dee D et al (2011) The ERA-interim reanalysis: configuration and performance of the data assimilation system. Q J Roy Meteorol Soc 137(656):553–597. https://doi.org/10.1002/qj.828/full

Dumitru A, Merticariu V, Baumann P (2014) Exploring cloud opportunities from an array database perspective. Proc ACM SIGMOD Workshop on Data Analytics in the Cloud (DanaC), Snowbird, USA, June 22–27, 2014

EarthServer (2015) EarthServer: big datacubes at your fingertips. www.earthserver.eu. Accessed 21 May 2017

ECMWF (2014) Data services leaflet 2014 - commercial license. www.ecmwf.int/sites/default/files/DataOservicesOleaflet020140commercialOlicence.pdf. Accessed 1 Aug 2016

ECMWF (n.d.) Web Coverage Service (WCS) for Climate Data. earthserver.ecmwf.int. Accessed 21 May 2017

GEOSS (2016) GEOSS resource details: rasdaman. geossregistries.info/geosspub/resource_details_ns.jsp?compId=urn:geoss:csr:resource:urn:uuid:e8d4eb7b-4681-4f71-9e34-0f78cf10ce92. Accessed 21 May 2017

Google (n.d.) Google Earth Engine. https://earthengine.google.com. Accessed 21 May 2017

Heather D, Barthelemy M, Manaud N, Martinez S, Szumlas M, Vazquez J, Osuna P (2013) ESA's planetary science archive: status, activities and plans. Eur Planet Sci Cong 8:EPSC2013-626

Hogan P (2011) NASA world wind: infrastructure for spatial data. Proc. 2nd International Conference on Computing for Geospatial Research & Applications. ACM, New York, NY

INSPIRE (2016) Technical Guidance for the Implementation of INSPIRE Download Services using Web Coverage Services (WCS). http://inspire.ec.europa.eu/id/document/tg/download-wcs. Accessed 21 May 2017

ISO (2004) Geographic information - schema for coverage geometry and functions

ISO (2015) ISO 9075 SQL Part 15: multi-dimensional arrays. ISO candidate standard, 2015

Masó J, Pomakis K Julià N (2010) OpenGIS Web Map Tile Service Implementation Standard, version 1.0. OGC document 07-057r7

McMahon S (1996) Overview of the planetary data system. Planet Space Sci 44(1):3–12

MEEO (2016) MEA multi-sensor evolution analysis. www.meeo.it/wp/products-and-services/mea-multisensor-evolution-analysis/. Accessed 21 May 2017

MEEO (2017) EO data service, eodataservice.org. Accessed 21 May 2017

Merticariu G, Misev D, Baumann P (2015) Measuring storage access performance in array databases. Proc. 7th Workshop on Big Data Benchmarking (WBDB), December 14–15, 2015, New Delhi, India

Misev D, Baumann P (2015) Enhancing science support in SQL. Workshop Data and Computational Science Technologies for Earth Science Research (co-located with IEEE Big Data), Santa Clara, US, October 29, 2015

NASA (2016) Web WorldWind. webworldwind.org. Accessed 21 May 2017

OGC (2016a) Coverages and web coverage service: the big picture. external.opengeospatial.org/twiki_public/pub/CoveragesDWG/CoveragesBigPicture/. Accessed 21 May 2017

OGC (2016b) WCS 2.0 (OGC 09-110r4) compliance test suite. cite.opengeospatial.org/teamengine/about/wcs/2.0.1/site/index.html. Accessed 21 May 2017

Oosthoek J, Flahaut J, Ross A, Baumann P, Misev D, Campalani P, Unnithan V (2013) PlanetServer: innovative approaches for the online analysis of hyperspectral satellite data from Mars. Adv Space Res 53(12):1858–1871

Paradigm4 (2016) SciDB reference manual: community and enterprise editions. Accessed 21 May 2017

PlanetServer (2016a) Planetary Science Data Service of EarthServer. www.planetserver.eu. Accessed 21 May 2017

PlanetServer (2016b) PlanetServer GitHub organization and repositories. github.com/planetserver. Accessed 21 May 2017

PML (2016) Marine data service. earthserver.pml.ac.uk. Accessed 21 May 2017

Pondrelli M, Tanaka K, Rossi A, Flamini E (eds) (2011) Geological mapping of Mars. Planet Space Sci. 59(11-12) 1113

rasdaman (2016a) Big Earth Data standards. standards.rasdaman.com. Accessed 21 May 2017

rasdaman (2016b) rasdaman. www.rasdaman.org. Accessed 21 May 2017

Rossi A, Oosthoek J, Baumann P, Beccati A, Cantini F, Misev D, Orosei R, Flahaut J, Campalani P, Unnithan V (2014) PlanetServer/EarthServer: big data analytics in planetary science. Geophys Res Abstr 16:EGU2014-5149

Sathyendranath S, Brewin R, Mueller D, Brockmann C, Deschamps P-Y, Doerffer R, Fomferra N, Franz BA, Grant M, Hu C, Krasemann H, Lee Z, Maritorena S, Devred E, Mélin F, Peters M, Smyth T, Steinmetz F, Swinton J, Werdell J, Regner P (2012) Ocean colour climate change initiative: approach and initial results. Geosci and Remote Sens 2012:2024–2027

Schut P (2007) Web processing service, version 1.0. OGC document 05-007r7, www.opengeospatial.org/standards/wps. Accessed 21 May 2017

Viviano-Beck CE et al (2014) Revised CRISM spectral parameters and summary products based on the currently detected mineral diversity on Mars. J Geophys Res Planet 119:1403–1431
Wikipedia (2016) Coverage data. en.wikipedia.org/Coverage_data. Accessed 21 May 2017
Zhang Y, Kersten M, Ivanova M, Nes N (2011) SciQL, bridging the gap between science and relational DBMS. Proc. IDEAS. ACM, New York, NY, pp 124–133

10

Development of an Earth Observation Cloud Platform in Support to Water Resources Monitoring

Andreea Bucur, Wolfgang Wagner, Stefano Elefante, Vahid Naeimi, and Christian Briese

Earth observation (EO) satellites collect verifiable observations that allow tracing natural and anthropogenic changes from local to global scale over several decades. Multi-decadal data sets are already available from various types of EO sensors, but their effective exploitation is hindered by the lack of data centres which offer dedicated EO processing chains and high-performance processing (HPC) capabilities. Recognizing this need, TU Wien founded the EODC Earth Observation Data Centre for Water Resources Monitoring together with other Austrian partners in May 2014 as a public–private partnership. The EODC aims at providing an independent science-driven platform that is transparent for its users and offering a high diversity and flexibility in terms of data sets and algorithms used. In this contribution, we describe the collaborative approach followed by EODC to build up its infrastructure and services and briefly introduce three pilot services.

A. Bucur (✉) • S. Elefante • V. Naeimi
Department of Geodesy and Geoinformation, Technische Universität Wien, Gußhausstraße 27-29/E120, 1040, Vienna, Austria
e-mail: andreea.bucur@geo.tuwien.ac.at; stefano.elefante@geo.tuwien.ac.at; vahid.naeimi@geo.tuwien.ac.at

W. Wagner
Department of Geodesy and Geoinformation, Technische Universität Wien, Gußhausstraße 27-29/E120, 1040, Vienna, Austria

EODC Earth Observation Data Centre for Water Resources Monitoring GmbH, Gusshausstrasse 27-29/CA 02 06, 1040, Vienna, Austria
e-mail: wolfgang.wagner@geo.tuwien.ac.at

C. Briese
EODC Earth Observation Data Centre for Water Resources Monitoring GmbH, Gusshausstrasse 27-29/CA 02 06, 1040, Vienna, Austria
e-mail: christian.briese@eodc.eu

Introduction

Humans have changed the natural environment since their early existence. However, the scale of human impacts has become dramatic only in the past 60 years (Steffen et al. 2015). During this so-called "Great Acceleration" period (McNeill and Engelke 2016) scientific and technological progress lead to extensive production and offer of goods and services, and an overall improvement in the standard of living for billions of people (Bhaduri et al. 2014). This period also witnessed a sharp increase in the world population, going up from three billion in 1959 to seven billion in 2012 (US Census Bureau 2016). All this had a dramatic impact on the consumption of natural resources. One of the resources, which are increasingly under pressure, is water. Water is pivotal for the well-being of humans and natural ecosystems: agricultural and industrial production, biodiversity, human health etc. In the "Global Risks 2015" report of the World Economic Forum, the "water crisis" is rated as the risk with the highest societal impact (World Economic Forum 2015). Therefore, it is crucial to understand natural and anthropogenic influences on the water cycle and the factors that might determine changes over time (e.g. Oki et al. 2004; Tang and Oki 2016). Major attention must be given to the rise in global temperature – e.g. year 2016 (January–October) was reported as the warmest in historical records (NOAA 2016)—and consequently to a warmer climate which is generally acknowledged to prompt an increased occurrence of extreme events such as floods and droughts (e.g. IPCC 2013; Trenberth and Asrar 2014).

In this context continuous monitoring of water resources is essential. In order to improve water management practices reliable information about anthropogenic and natural impacts, and their interactions must be readily available. Ground-measurements are fundamental for this purpose. However, they have many short-comings: sparse information over small areas, lack of representativeness at larger scale, high costs of maintenance, out of date or failed equipment and lack of funds to replace them etc. Complementing in situ networks, monitoring tasks are increasingly fulfilled by earth observation (EO) satellites which have been acquiring measurements of the land, atmosphere and oceans since the beginning of the 1970s. The new generation of sensors is able to collect an unprecedented amount and variety of observational data at high spatial resolution and short repeat intervals. A vast and diverse amount of EO data is, therefore, readily available to be mined for new insightful information; but this task is not short of challenges. As we will detail further in section "EODC: The Earth Observation Data Centre for Water Resources Management", dedicated data centres that stimulate collaboration are needed for the effective exploitation of satellite images.

Here we present the EODC Earth Observation Data Centre for Water Resources Management which was founded as a public–private partnership with the aim to assist in water management by making use of earth observation data and big data cloud computing infrastructures. In the next section, the organisational and technical aspects of EODC are presented. In section "Pilot Services", initial pilot services are briefly described.

EODC: The Earth Observation Data Centre for Water Resources Management

The EODC Earth Observation Data Centre for Water Resources Monitoring (www.eodc.eu) is a public-private partnership founded in May 2014, in Austria, by the Technische Universität Wien (TU Wien), the Austrian Meteorological and Geodynamics Institute (ZAMG), two private companies and individuals. The early idea of EODC was born already in 2011, and was prompted by the need to cope with exponentially growing data volumes and their scientific exploitation with increasingly complex algorithms (Wagner et al. 2014). EODC was set up as an international cooperation network which brings together scientific institutions, public organizations and several private partners from countries within and outside Europe.

Working with EO data on cloud platforms is not short of scientific, technical and organizational challenges as described in Wagner et al. (2014). The science is driven by the need to gain an integrated view of all processes driving the water cycle (Wagner et al. 2009). This requires analyses of many different geophysical parameters and their coupled feedbacks (e.g. soil moisture, temperature, precipitation, vegetation indices) based on data from multiple sensors (e.g. active and passive radar, optical imaging satellites) and their integration into earth system models. Thus, the information contained in satellite images becomes meaningful only after several specific processing steps.

Traditionally, the ground segments of EO missions have delivered raw images to remote sensing experts who, after high-level data processing (geo-referencing, normalization, radiometric correction etc. of data), have handed it out to application oriented users (hydrology, forestry, urban planning etc.). The later have extracted added-value information which can be further used for specific purposes (mapping for forest management etc.). This long-established system has assured that all parties had full control over the ownership of data and software. But this approach is inefficient because the data and resources (such as storage and processing capabilities or specific expertise for EO data processing) are basically duplicated for each user. Today, this traditional approach is reaching its limits. This is because, firstly, the latest generation of sensors generate huge amounts of data. To give one example, European Space Agency's Sentinel-1 satellites acquire in one year more data than their predecessor ENVISAT Advanced Synthetic Aperture Radar (ASAR) has done so in 10 years of operation (25 Terabytes in the first year of S1, 23.5 Terabytes in 10 years of ASAR). With a data capture rate of about 1.8 Terabytes per day, Sentinel-1 will acquire over its 7-year nominal mission lifetime over 1 Petabyte of raw data (Wagner 2015). Secondly, the algorithms used to transform the

EO data into useful information become increasingly more complex. Last but not least, a model may be run with more than just one data set or several complementary methods are combined into an ensemble in order to obtain the most reliable results and to estimate the uncertainty range of the predictions.

Considering the above, the way how EO data are stored, processed and distributed needs to be changed fundamentally. This has already been recognized by a number of private and public entities that have started to offer big data infrastructures for processing EO data (Wagner 2015). Some examples include private companies such as Google, and Amazon, and the public initiatives THEIA Land Data Centre in France or the Climate, Environment and Monitoring from Space (CEMS) initiative in the UK. Their solutions typically combine cloud technologies and high-performance computing (HPC) to allow users to explore large amounts of data via an internet connection. In other words, "the software moves to the data" rather than data being moved to the software on local working stations.

Similar to the above-mentioned entities also the EODC offers such a novel framework for working with EO data; its users have the possibility to access EO data via a cloud platform, process them with their own algorithms and extract the results. In order to be attractive for both scientific and operational users, the EODC infrastructure combines elements for operational data reception and processing, cloud platforms and storage system for scientific analysis collocated with advanced HPC capabilities. The EODC infrastructure has not been built from scratch but exploits as good as possible existing data centre capabilities by federating and integrating them. The main EODC data centre capabilities are currently located at the TU Wien Science Centre Arsenal collocated with the Vienna Scientific Cluster 3 (VSC-3). Figure 1 gives an overview of the status of this infrastructure at the end of 2016. It is planned to successively extend these capabilities in order to allow storing and processing the complete global Sentinel data archives.

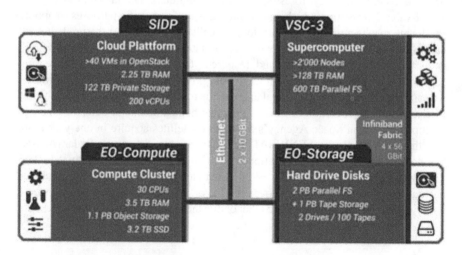

Fig. 1 The EODC infrastructure at the end of 2016

Several experiments carried out with Sentinel-1 SAR and ENVISAR ASAR data sets have already demonstrated the scalability of the EODC supercomputing environment. For example, a batch of 31, 978 Sentinel-1 images over Europe, with a total size of around 30 Terabytes (TB), was processed with TU Wien's SAR Geophysical parameters Retrieval Toolbox (SGRT). This Python package incorporates the ESA's Sentinel-1 Toolbox (S1TBX) and consists of modules for EO data pre-processing, model parameters extraction, and data production (Naeimi et al. 2016). Processing the 31, 978 Sentinel-1 images on the VSC-3 with around 300 nodes took roughly 10 days compared to more than 1 year that would be needed when processing the same data set with the same software with only 1 node (Elefante et al. 2016).

In terms of data availability, EODC hosts at the TU Wien Science Centre Arsenal a nearly complete and up-to-date data archive from its main sensors of interest (Sentinel-1, Sentinel-2, Sentinel-3). Additional data are available through the other EODC data centres operated by EODC cooperation partners (ZAMG, VITO NV, EURAC research). In this way the EODC decentralised IT infrastructure provides its users access to an extended and diverse number of data sets, trying to minimise the duplication of data as much as possible.

An ultimate goal of EODC is to encourage its partners and users to engage in collaborative science activities. The organisational structure was design to facilitate this by offering more than just access to performant processing resources. Thus, as described in Wagner et al. (2014), partners come together in so called communities which are formed around particular research topics (e.g. soil moisture), applications/services (e.g. drought monitoring, flood mapping) or tasks (e.g. software development, shared infrastructure resources). The participation in the EODC cooperation network is flexible according to one's interests and contribution, and can take one of the three forms of partnership: Principal Cooperation Partners, Associated Cooperation Partners or Developers. Facilitated by this bundling of interests, several EO data services are currently being developed jointly by several EODC partners.

Pilot Services

Several joint EODC services are already under development. These services typically rely on individual sensors, but ultimately the goal will be to benefit from the collocation of many diverse data sets by building multi-sensor data services. An example for single-sensor service is the Sentinel-2 data service platform developed by researches of the University of Natural Resources and Life Sciences (BOKU), Vienna, and run on the EODC infrastructure. As described by Vuolo et al. (2016), users of this service platform can submit processing requests and access the results via a user-friendly web page or using a dedicated application programming interface (API). Data products that can be produced in this way are atmospherically corrected

Sentinel-2 images and value-added products with a particular focus on agricultural vegetation monitoring, such as leaf area index (LAI) and broadband hemispherical-directional reflectance factor (HDRF).

An example for a multi-sensor service is the ESA CCI soil moisture data service as descried by Dorigo et al. (2017). Soil moisture is an important component of the water cycle and the satellite-based products derived from active and passive microwave are increasingly being used for a wide range of applications (Dorigo and de Jeu 2016). For example, satellite-based soil moisture may be used for estimation of near-future vegetation health (Qiu et al. 2014), improved calculation of crop water requirement (McNelly et al. 2015) and operational drought warnings (Enenkel et al. 2016). EODC currently leads the second phase of the ESA Climate Change Initiative Soil Moisture project, providing the operational framework for merging more than a dozen of satellite data sets into consistent long-term soil moisture data records (Liu et al. 2012, 2011; Wagner et al. 2012).

As a last example, we note that several Sentinel-1 data services are currently being developed by the Remote Sensing Research Group of TU Wien in collaboration with other EODC partners. They range from a simple Sentinel-1 image-compositing service to processing services for the monitoring of soil moisture, water bodies, wetlands, and forests from regional to global scales. Figure 2 illustrates a false-colour composite of the Sentinel-1 data acquisitions before and after a flooding event on December 2015 (BBC News 2015) in Carlisle, UK.

Closing Remarks

In this subchapter we introduced the EODC Earth Observation Data Centre for Water Resources Management, which is a private–public entity founded for enabling the collaboration of scientific, public and private organizations for processing EO data in the cloud. As its name suggests, one of founding idea of EODC was to focus on the thematic area of water resources management, but thanks to the rapid growth of the EODC cooperation network, the number of application domains has been growing accordingly. In particular, agricultural monitoring and land use mapping applications have become important topics of collaboration between EODC partners. The experiences made over the short period since the foundation of EODC in 2014 show that EODC offers a framework for collaboration that can assist the development of long and complex data processing lines going from the raw EO data to the final model predictions (runoff forecast, crop yield etc.). Taking advantage of the big data technologies, latest scientific algorithms can be scaled up to process high-resolution EO data from regional to global scales. This is a crucial step towards operational applications, which are ultimately needed to enhance the social benefits of EO technology.

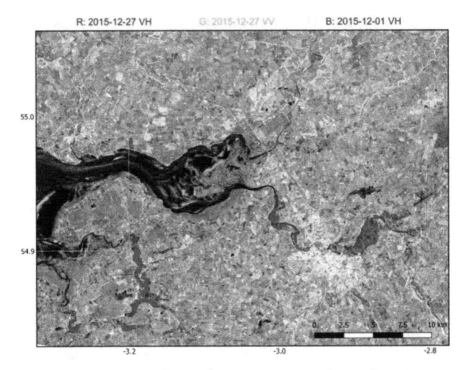

Fig. 2 False-colour composite generated from Sentinel-1 data acquisitions in Interferometric Wide swath (IW) mode, Ground Range Detected (GRD) high resolution product (20 m), VV and VH polarization from 1st and 27th of December 2015

References

BBC News (2015) December storms' trail of destruction, 29 Dec 2015. Available from: http://www.bbc.com/news/uk/35193682. Accessed on 10 Dec 2016

Bhaduri A, Bogardi J, Leentvaar J, Marx S (2014) The global water system in the anthropocene. Challenges for science and governance. Springer International Publishing, New York, NY. 437 pages

Dorigo W, Wagner W, Albergel C, Albrecht F, Balsamo G, Brocca L, Chung D, Ertl M, Forkel M, Gruber A, Haas E, Hamer P, Hirschi M, Ikonen J, de Jeu R, Kidd R, Lahoz W, Liu YY, Miralles D, Mistelbauer T, Nicolai-Shaw N, Parinussa R, Pratola C, Reimer C, van der Schalie R, Seneviratne SI, Smo-lander T, Lecomte P (2017) ESA CCI soil moisture for improved Earth system understanding: state-of-the art and future directions. Remote Sens Environ. in press. https://doi.org/10.1016/j.rse.2017.07.001

Dorigo W, de Jeu R (2016) Satellite soil moisture for advancing our understanding of earth system processes and climate change. Int J Appl Earth Observ Geoinform 48:1–4

Elefante S, Wagner W, Briese C, Cao S, Naeimi V (2016) High-performance computing for soil moisture estimation. In Proceedings of the 2016 conference on Big Data from Space (BiDS'16), Santa Cruz de Tenerife, Spain, pp 95–98. https://doi.org/10.2788/854791

Enenkel M, Steiner C, Mistelbauer T, Dorigo W, Wagner W, See L, Atzberger C, Schneider S, Rogenhofer E (2016) A combined satellite-derived drought indicator to support humanitarian aid organizations. Remote Sens 8(4):340

IPCC (2013) Climate change 2013: the physical science basis. In: Stocker TF, Qin D, Plattner G-K, Tignor M, Allen SK, Boschung J, Nauels A, Xia Y, Bex V, Midgley PM (eds) Contribution of Working Group I to the Fifth Assessment Report of the Intergovernmental Panel on climate change. Cambridge University Press, Cambridge., 1535 pp. https://doi.org/10.1017/CBO9781107415324

Liu YY, Dorigo WA, Parinussa RM, De Jeu RAM, Wagner W, McCabe MF, Evans JP, Van Dijk AIJM (2012) Trend-preserving blending of passive and active microwave soil moisture retrievals. Remote Sens Environ 123:280–297

Liu YY, Parinussa RM, Dorigo WA, De Jeu RAM, Wagner W, Van Dijk AIJM, McCabe MF, Evans JP (2011) Developing an improved soil moisture dataset by blending passive and active microwave satellite-based retrievals. Hydrol Earth Syst Sci 15:425–436

McNeill JR, Engelke P (2016) The great acceleration: an environmental history of the anthropocene since 1945. Harvard University Press, Cambridge, MA

McNelly A, Husak GJ, Brown M, Carroll M, Funk C, Yatheendradas S, Arsenault K, Peters-Lidard C, Verdin JP (2015) Calculating crop water requirement satisfaction in the West Africa Sahel with remotely sensed soil moisture. J Hydrometeorol 16(1):295

Naeimi V, Elefante S, Cao S, Wagner W, Dostalova A, Bauer-Marschallinger B (2016) Geophysical parameters retrieval from sentinel-1 SAR data: a case study for high performance computing at EODC. In Proceedings of the 24th High Performance Computing Symposium (HPC '16). Society for Computer Simulation International, San Diego, CA., Article 10, 8 pages. 10.22360/SpringSim.2016.HPC.026

NOAA (2016) National Centers for Environmental Information, State of the Climate: Global Analysis for October 2016 (published online November 2016). Available from: http://www.ncdc.noaa.gov/sotc/global/201610. Accessed on 10 Dec 2016

Oki T, Entekhabi D, Harrold TI (2004) The global water cycle. In: The state of the planet: Frontiers and challenges in Geophysics. American Geophysical Union, Washington, DC, pp 225–237

Qiu J, Crow WT, Nearing GS, Mo X, Liu S (2014) The impact of vertical measurement depth on the information content of soil moisture times series data. Geophys Res Lett., 2014GL060017. https://doi.org/10.1002/2014GL060017

Steffen W, Broadgate W, Deutsch L, Gaffney O, Ludwig C (2015) The trajectory of the Anthropocene: the great acceleration. Anthrop Rev 2(1):81–98

Tang Q, Oki T (eds) (2016) Terrestrial water cycle and climate change: natural and human-induced impacts, vol 221. John Wiley & Sons., New York, NY

Trenberth KE, Asrar GR (2014) Challenges and opportunities in water cycle research: WCRP contributions. Surv Geophys 35(3):515–532

Vuolo F, Żółtak M, Pipitone C, Zappa L, Wenng H, Immitzer M, Weiss M, Baret F, Atzberger C (2016) Data service platform for Sentinel-2 surface reflectance and value-added products: System use and examples. Remote Sens 8:938

US Census Bureau (2016) International Data Base, Updated August 2016. Available from: www.census.gov. Accessed on 10 Dec 2016

Wagner W (2015) Big data infrastructures for processing sentinel data. In: Fritsch D (ed) Photogrammetric week, pp 93–104

Wagner W, Fröhlich J, Wotawa G, Stowasser R, Staudinger M, Hoffmann C, Walli A, Federspiel C, Aspetsberger M, Atzberger C, Briese C, Notarnicola C, Zebisch M, Boresch A, Enenkel M, Kidd R, von Beringe A, Hasenauer S, Naeimi V, Mücke W (2014) Addressing grand challenges in earth observation science: the Earth Observation Data Centre for water resources monitoring. ISPRS Annals Photogram Remote Sens Spatial Inform Sci 2(7):81

Wagner W, Dorigo W, de Jeu R, Fernandez-Prieto D, Benveniste J, Haas E, Ertl M (2012) Fusion of active and passive microwave observations to create an Essential Climate Variable data record on soil moisture. In: XXII ISPRS Congress, Melbourne, Australia

Wagner W, Verhoest N, Ludwig R, Tedesco M (2009) Remote sensing in hydrological sciences. Hydrol Earth Syst Sci 13(6):813–817

World Economic Forum (2015) Global risks 2015, 10th edn. World Economic Forum, Geneva

11

Cyber-Infrastructure for Data-Intensive Geospatial Computing

Rajasekar Karthik, Alexandre Sorokine, Dilip R. Patlolla, Cheng Liu, Shweta M. Gupte, and Budhendra L. Bhaduri

With the recent advent of heterogeneous High-performance Computing (HPC) to handle EO "Big Data" workloads, there is a need for a unified Cyber-infrastructure (CI) platform that can bridge the best of many HPC worlds. In this chapter, we discuss such a CI platform being developed at Geographic Information Science and Technology (GIST) group using novel and innovative techniques, and emerging technologies that are scalable to large-scale supercomputers. The CI platform utilizes a wide variety of computing such as GPGPU, distributed, real-time and cluster computing, which are being brought together architecturally to enable data-driven analysis, scientific understanding of earth system models, and research collaboration. This development addresses the need for close integration of EO and other geospatial information in the face of growing volumes of the data, and facilitates spatio-temporal analysis of disparate and dynamic data streams. Horizontal scalability and linear throughput are supported in the heart of the platform itself. It is being used to support very broad application areas, ranging from high-resolution settlement mapping, national bioenergy infrastructure to urban information and mobility systems. The platform provides spatio-temporal decision support capabilities in planning, policy and operational missions for US federal agencies. Also, the platform is designed to be functionally and technologically sustainable for continued support of the US energy and environment mission for the coming decades.

R. Karthik (✉) • A. Sorokine • D.R. Patlolla • C. Liu • S.M. Gupte • B.L. Bhaduri
Geographic Information Science and Technology Group, Oak Ridge National Laboratory, Oak Ridge, TN, USA
e-mail: karthikr@ornl.gov; sorokina@ornl.gov; patlolladr@ornl.gov; liuc@ornl.gov; guptesm@ornl.gov; bhaduribl@ornl.gov

Introduction

In today's world, there is a multitude of heterogeneous Earth Observation High-performance Computing (EO-HPC) systems, each designed to solve specific science or technological missions. These EO-HPC systems often work independently of each other, hindering the flow of information and limiting the ability to achieve interoperability among systems. Bringing these systems together to create a fully and closely knitted Cyber-infrastructure (CI) platform provides a shared universe for EO data driven analysis and discovery capabilities for US federal agencies. Though there have been some promising community efforts to build a CI platform that can integrate various types of EO-HPC systems together, there does not exist a unified CI platform currently (Kalidindi 2015; Bhaduri et al. 2015a). With "Big Data" explosion changing the landscape of software architecture design, there is an increasing need for such a platform that can meet the modern complex needs.

Geographic Information Science and Technology (GIST) and The Urban Dynamics Institute (UDI) at Oak Ridge National Laboratory (ORNL) are building such a CI platform by integrating our next-generation EO-HPC systems under one umbrella using novel techniques and emerging technologies. We employ wide breadth of techniques across the spectrum of scientific computing such as GPGPU, distributed, real-time and cluster computing (Bhaduri et al. 2015a; Karthik 2014a; Sorokine et al. 2012). The platform is designed to scale to petabytes of data and handle massive workloads. One of the biggest architectural challenges in developing our CI platform was addressing complex interdependencies among the various systems without compromise in efficiency or functionalities of individual systems. The platform achieves foundational, structural and semantic interoperabilities to create a harmonized and seamless experience across various systems. With this platform, various systems are designed to work together as a whole information system, but also retain the ability to operate independently if desired. In this integrated platform, the systems communicate with each other providing various levels of control of components and functionality, yet allowing for independent control as well. Modularity is being designed at the core of the CI platform to support future EO-HPC systems for easier integration, and foster sustainable development to meet US science and energy missions now and into the future. Our CI platform aims to take the next major leap in Data Science and Cyber-infrastructure, while making the best use of our existing EO-HPC systems.

In the following sections, we describe the challenges, trends and how our CI platform plays an important role in our research initiatives illustrated with settlement mapping, mobility science, and urban information system.

Settlement Mapping Tool (SMTOOL)

Understanding high-resolution population distribution data is fundamental to reduce disaster risk, eliminate poverty, and foster sustainable development. Modern censuses fail to cover population in remote, inaccessible areas in many underdeveloped nations. Small settlements are visible in high-resolution satellite imagery, which have historically been computationally expensive for information extraction. Utilizing GPUs, Oak Ridge National Laboratory is mapping the smallest settlements and associated population across the globe for the first time in our history. The high-resolution settlement and resulting 90 m LandScan HD population data have profoundly enhanced our ability to reach and serve vast vulnerable populations from local to planet scales (Bhaduri et al. 2015b).

Past 50 years have witnessed the global population increase by four billion and with 150 new births every minute, an additional four billion people will settle on this planet in the coming 50. Urban and rural population distribution data are fundamental to prevent and reduce disaster risk, eliminate poverty and foster sustainable development. Commonly available population data, collected through modern censuses, do not capture this high-resolution population distribution and dynamics. However, there is gross underestimation of global human settlements and population distribution. Footprints of our expanding activities are impacting the future of this planet from availability of natural resources to a changing climate. Accurate assessment of high-resolution population data is essential for successfully addressing key issues such as good governance, poverty reduction strategies, and prosperity in social, economic and environmental health. Geospatial data and models offer novel approaches to disaggregate Census data to finer spatial units; with land use and land cover (LULC) data being the primary driver. With increasing availability of LULC data from satellite remote sensing, "developed" pixels have been nucleus to assessing settlement build up from human activity. However, the processing and analysis of tera to peta scale satellite data has been computationally expensive and challenging.

With the availability of moderate to high resolution LULC data derived from NASA MODIS (250–500 m) or Landsat TM (30 m) have facilitated the development of population distribution data at a higher spatial resolution such as Oak Ridge National Laboratory's (ORNL) LandScan Global (1 km) and LandScan USA (90 m); two finest resolution population distribution data developed. Although these LULC data sets have somewhat alleviated the difficulty for population distribution models, in order to assess the true magnitude and extent of the human footprint, it is critical to understand the distribution and relationships of the small and medium-sized human settlements. These structures remain mostly undetectable from medium resolution satellite derived LULC data. For humanitarian missions, the truly vulnerable, such as those living in refugee camps, informal settlements and slums need to be effectively and comprehensively captured in our global understanding. This is particularly true in suburban and rural areas, where the population is dispersed to a greater degree than in urban areas. Extracting settlement information from very high-resolution (1 m or finer), peta-scale earth observation imagery has been a promising pathway for rapid estimation and revision of settlement and population distribution data. As early as 2005, automated feature extraction algorithms implemented on available CPU-based architectures demonstrated radical improvement in image analysis efficiency when manual settlement identification from a 100-km^2 area was reduced from 10 h to 30 min. However, this scaled inefficiently with limited resources on a workstation and at that rate processing 57-million km^2 habitable area would take decades. The pressing need for identifying population distribution in the smallest human settlements and monitoring settlement patterns at local to planet scale as landscape changes are induced by population growth, migration, and disasters, compels a computational solution to process large volumes of very high resolution satellite imagery. Such a solution did not exist before this work was accomplished at ORNL.

Extracting settlements at high-resolution satellite images was accomplished by mapping sub-meter pixel data to unique patterns that correlate with the underlying settlements. To account for the variations in settlement structures spanning from skyscrapers to small dwellings, we generated patterns at different scales. The mapping process involved simultaneously analyzing a set of connected pixels to extract low-level structural features such as building edges, corners, and lines. Next, the spatial arrangements of these structural features at different scales were computed in parallel allowing us to generate unique settlement signatures efficiently. Furthermore, pattern recognition based on a previously learned model was integrated with the pattern generation step allowing us overcome the need to for additional storage. Our strategy of mapping pixels to underlying structural patterns was quite different from existing approaches that relied on spectral measurements such as reflectance at different wavelengths for settlement detection.

This approach using sub-meter resolution imagery is not only useful in generating accurate human settlement maps, but also it allows potential (social and vulnerability) characterization of population from settlement structures (tents, huts, buildings) from image texture and spectral features. Rapid ingestion and analysis of high resolution imagery to enhance quality and timely availability of input spatial data provides a cost and time effective solution for developing current and accurate high resolution population data. Such progresses in geospatial science and technology hold tremendous promise for advancing the state of accuracy and timely flow of critical geospatial information not only to benefit numerous sustainable development programs; but also has significant implications for time critical missions of disaster support.

High-resolution settlement data is foundational information for locating populations and activities in an area. One major usage of this data is as input to ORNL's LandScan HD population distribution model, which combines the settlement data with population density information to generate population distribution at an unprecedented 90 m resolution. For many underdeveloped countries, official censuses never reach remote, difficult to access areas. Moreover, there have not been reasons to exploit high-resolution satellite imagery for those areas. Consequently, this capability has enabled us to locate populations in secluded areas of the planet for the first time in our history. High-resolution settlement and population data are being used by the global humanitarian community for missions ranging from planning critical infrastructure and services to the deserving population, responding to the Ebola crisis in western Africa, eradicating Polio in Nigeria, as well as defining and mitigating disaster risks.

Fig. 1 An overview of
SMTOOL architecture

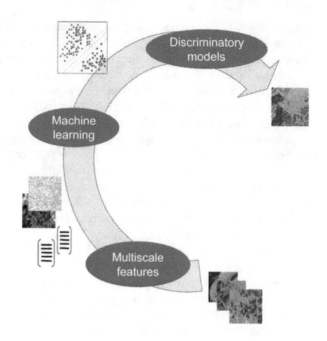

Often solutions require advanced algorithms capable of extracting, representing, modeling, and interpreting scene features that characterize the spatial, structural and semantic attributes. Furthermore, these solutions should be scalable enabling analysis of big image datasets; at half-meter pixel resolution the earth's surface has roughly 600 Trillion pixels and the requirement to process at this scale at repeated intervals demands highly scalable solutions. Thus, we developed a GPU-based computational framework (as illustrated in Fig. 1) designed for identifying critical infrastructures from large-scale satellite or aerial imagery to assess vulnerable population. We exploit the parallel processing capability of GPUs to present GPU-friendly algorithms for robust and efficient detection of settlements from large-scale high-resolution satellite imagery (Patlolla et al. 2012). Feature descriptor generation is an expensive (computationally demanding), but a key step in automated scene analysis. To address the large-scale data processing needs we exploited the parallel computing architecture and carefully designed our algorithm to scale with hardware and fully utilize the memory bandwidth (required to transfer the high resolution image data) efficiently to produce great speedups times for the feature descriptor computation (as illustrated in Fig. 2).

We could thus achieve GPU-based high speed computation of multiple feature descriptors—multiscale Histogram of Oriented Gradients (HOG) (Patlolla et al. 2012), Gray Level Co-Occurrence Matrix (GLCM) Contrast, local pixel intensity statistics, Texture response (local Texton responses to a set of oriented filters at each pixel) (Patlolla et al. 2015), Dense Scale Invariant Feature Transform (DSIFT), Vegetation Indices (NDVI), Line Support Regions (extraction of straight line segments from an image by grouping spatially contiguous pixels with consistent

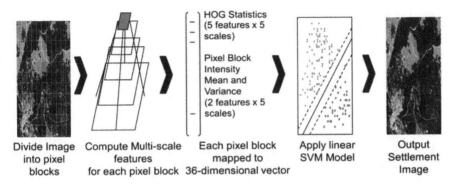

Fig. 2 Settlement mapping process

orientations), Band Ratios (a digital image-processing technique that enhances contrast between features by dividing a measure of reflectance for the pixels in one image band by the measure of reflectance for the pixels in the other image band) etc. Once, the features are computed, a linear SVM is used to classify settlement and non-settlements. The computational process requires dozens of floating point computations per pixel, which can result in slow runtime even for the fastest of CPUs. The slow speed of a CPU is a serious hindrance to productivity for time critical missions. Our GPU-accelerated computing solution provides an order of magnitude or more in performance by offloading compute-intensive portions of the application to the GPU, while the remainder of the code still runs on the CPU. The implementation further scales linearly with the available nodes thus enabling the processing of large-scale data on high end GPU-based cluster-computers.

With the introduction of very high-resolution satellite imagery, mapping of small or spectrally indistinct settlements became possible on a global scale. However, existing methodologies for extracting and characterizing settlements rely on manual image interpretation or involve computationally intensive object extraction and characterization algorithms that saturate the computational capabilities of conventional CPU-based options used by commercial remote sensing software packages. Many of these existing pixel based image analysis techniques used for medium resolution Landsat imagery ($\sim$30 m) or coarser MODIS imagery (250–1000 m) are not ideal for interpreting satellite imagery with sub-meter spatial resolution. Advanced modeling of the spatial context is necessary to extract and represent information from such high-resolution overhead imagery. On a global scale, critical computational challenges are posed in the processing of petabytes of sub-meter resolution. For example an image of Kano city in Nigeria at 0.5 m resolution represents 23 Gigabytes of data covering 13,050 km^2. Attempts to accurately extract settlements using CPU-based commercial remote sensing packages were unsuccessful. An estimate to manually digitize the settlements for this area was 870 h. Meanwhile, SMTool on a 4 Tesla GPU workstation is able to process this large dataset in approximately 17 min (as illustrated in Table 1, Figs. 3 and 4).

Table 1 Performance of various features—processing times are based on a 4 C2075 GPU workstation

Feature	Accuracy (%)	Runtime (s)
HOG	93.5	1.6
TEXTONS	92.7	4.7
VEGIND	91.4	1.77
BANDRT	86.1	1.93
DSIFT	90.8	11.33

Fig. 3 TEXTONS performance—33 images of 0.5 m spatial resolution, each covering an area of 2.6 km^2, collected from various parts of Kandahar, Afghanistan

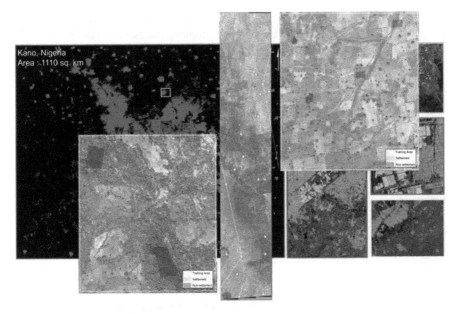

Fig. 4 SMTOOL results for Kano, Nigeria—On an average ~1% training samples

The Compute Unified Device Architecture (CUDA) has enabled us to efficiently utilize NVIDIA TESLA GPUs to develop and utilize (in a practical and efficient manner) the expensive feature descriptor algorithms that would otherwise be complex and impractical.

It is quite important to carefully design the computing strategies to fully exploit GPU's parallel computational capabilities. For this we divide the high-res imagery into non-overlapping square pixel-blocks consisting of $M \times M$ pixels. To set an optimal value for M, we experimented with several values ranging from 8 to 64 pixels. For each of the pixel-block we compute feature descriptors at several scales based on the low-level image features. Parallel processing is key to efficiently implementing the feature descriptor algorithm, where thousands of threads can run simultaneously. The K20X Tesla GPU in the GEOCLOUD cluster has 2688 CUDA cores/GPU at ~3.95 Tera Flops of single precision computing power, which helped us deliver speedups ranging from 100× to 220×, thus cutting the feature descriptor computation time from days to mere minutes.

Efficient memory utilization is key for the optimal performance of our algorithm on the GPU as it involves transfer of large amounts of image raster data to the GPU and the output settlement layer back to the host. We leveraged the Tesla GPUs PCI-E 3.0 interface to achieve over 10 GB/s transfers between the host (CPUs) and the devices (GPUs). An important factor to consider is the data transfer between CPU and GPU and optimal speedups require reducing the amount of data transferred between the two architectures. In our implementation the data transfer between CPU and GPU is performed only at the beginning and the end of the process. First, the image is read using GDAL to store the data in the CUDA global memory, which is 6 GB and provides a 288 GB/s memory–memory bandwidth. Though the data requests to the global memory has higher latency, CUDA provides a number of additional methods for accessing memory (i.e. shared, constant memory etc.) that can remove the majority of data requests to the global memory, thus enabling us to keep the Compute to Global Memory Access (CGMA) ratio at high values to achieve fine grained parallelism.

This has been an interdisciplinary effort with team members with academic and professional training in geography, electrical engineering, computer science and engineering and expertise in geographical sciences, population and settlement geography, spatial modeling, satellite remote sensing, machine learning data analysis and high performance computing.

Toolbox for Urban Mobility Simulations (TUMS)

TUMS, Toolbox for Urban Mobility Simulations, is a web-based high-resolution quick response traffic simulation modeling system for urban transportation studies. TUMS can be used both as a daily commuter traffic simulator or an emergency evacuation planning tools. There are some unique features in TUMS comparing with other similar transportation modeling and traffic simulation systems. It uses high-resolution population distribution and detailed street network, both covering the entire world. TUMS is aiming to simulate county level traffic flow using microscopic traffic simulation modeling and it has web applications based on WebGL. Users can take advantage of client side Geographic Process Unit (GPU) when it presents on client desktop. The transportation engine of TUMS is based on an open source package called TRANSIMS (TRansportation ANalysis SIMulation System, version 5.0) (Smith et al. 1995).

Global Dataset

Two main datasets used in TUMS are a population distribution dataset called LandScan developed by ORNL (Bhaduri et al. 2002) and a worldwide open source street level transportation network, called OSM, OpenStreetMap (OpenStreetMap 2016). Both datasets covers the entire planet.

LandScan has two components, LandScanUSA and LandScanGlobal. As the name indicates, LandScanUSA is the population distribution for USA and LandScanGlobal covers the entire world including USA. Both dataset are updated yearly. LandScan divides the study area into cells and each cell has a population count. LandScanUSA has higher resolution cells than LandScanGlobal. LandScanUSA uses 3 arc second cells while LandScanGlobal uses 30 arc second cells. Roughly, the 3 arc second cell has the size of 90 m by 90 m and the 30 arc second cell has the size of 1 km by 1 km around the equator. The size of cells becomes smaller when the latitude is higher. In order to make the analysis consistent, TUMS decomposes LandScanGlobal to 3 arc second cells using a primitive moving average method. If the study area is within USA then TUMS uses LandScanUSA dataset. If the study area is outside of USA then TUMS uses the decomposed LandScanGlobal 3 arc second dataset. From now on in this paper, we will use LandScan to represent both LandScanUSA and decomposed LandScanGlobal with 3 arc second cells.

OSM is updated weekly. The data quality in OSM depends on geographic region. Europe and North America data has much higher quality than Asia and Africa. Since OSM keeps evolving, the data quality now has improved tremendous compared to earlier version. Please do not be conceived by its name, OpenStreetMap, it not only have street network, it has other features such as land use type, administration boundary, physical features and lots more. However, TUMS only uses street network at current stage.

Framework

There are three major components in TUMS framework, a pre-processing component, a traffic simulation component, and a web-based visualization component (Fig. 5). The pre-processing component is responsible for preparing the input data for the transportation modeling. This first step is to define a study area, which can be a county (in USA only), a polygon or a circle. The next step is to extract the population and street network from LandScan and OSM. After integrated these two

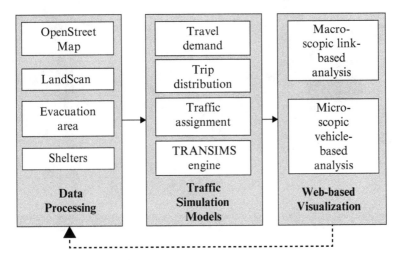

Fig. 5 TUMS framework

data together, TUMS creates a routable network with correct network topology and generates origin-destination (OD) tables for transportation modeling.

The traffic simulation models are based on TRANSIMS framework. TRANSIMS has more than a dozen executable programs, which are loosely coupled. Each executable program can be executed separately if the input data is properly set. Roughly, these programs can be grouped into five categories, synthetic population generation, network preparation, origin-destination (OD) table preparation, trip distribution and assignment, and Microscopic traffic simulation. TUMS takes the advantages of this flexible framework and integrate its own modules into TRANSIMS framework. For example, the synthetic population generation modules are replaced by LandScan population module. TUMS own OD table generation modules using LandScan and OSM substituted TRANSIMS OD table preparation modules.

Since TRANSIMS does not have a Graphic User Interface (GUI), TUMS developed two independent visualization tools for different background users (Karthik 2014b). The link based visualization and analysis tools are for planners who are interest in the measure of efficacy (MOE) for the planning purpose. The vehicle base animation tools are for traffic engineers who are more interested on operations such as intersection traffic control. Figures 6 and 7 are the examples for link-based and vehicle based GUI.

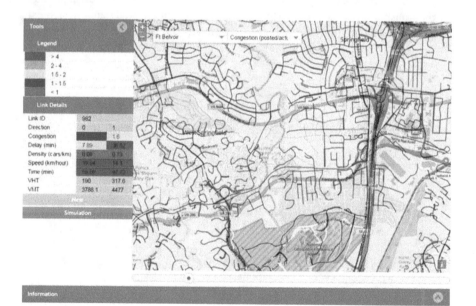

Fig. 6 Link-based visualization tool

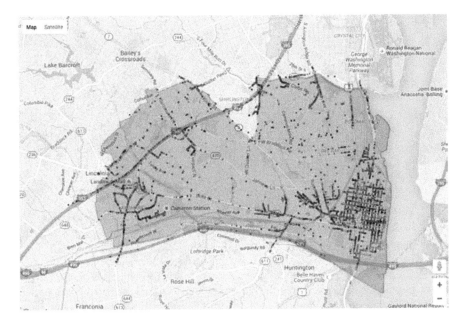

Fig. 7 Vehicle-based visualization tool

OD Tables

By manipulating the OD tables TUMS can simulate both daily commuter traffic flow or non-notice emergency evacuation simulation. Although LandScan only reports the total population count for each cell, but internally, LandScan has five layers, which are worker, residential, school, shopping and non-movement group. With these layers it is possible to generate the O-D tables for daily commuter traffic flow. Figure 8 is an example of daily commuter traffic flows for the year of 2015 and 2035.

For non-notice emergency evacuation, TUMS assumes that every evacuee would like to take a trip to the nearest shelter or exit point (boundary points) to get out of the evacuation area as quickly as possible. The OD tables for non-noticed emergency evacuation simulation are generated by finding the nearest shelters for each LPC.

Resolution

Traditional transportation models, both macroscopic and microscopic, use Traffic Analysis Zone (TAZ) for the OD tables. TAZs are the basic geographic unit for demographic data and land use type. The size of TAZ varies. Zones are smaller in urban area with high population density and larger in rural area with

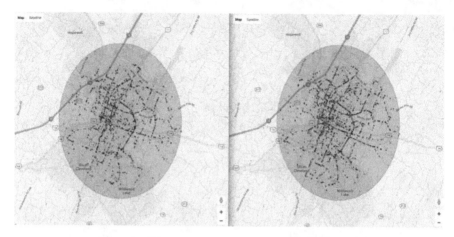

Fig. 8 Daily traffic flow simulation at Cleveland, TN, on the year 2015 (*Left*) and 2035 (*Right*)

lower population density. With the rising of agent-based and driver behaviour traffic modeling in transportation research, the large area TAZ is not suitable for microscopic traffic simulations. For example, Alexandria County, VA, has only 62 TAZs. Each TAZ covers quit large area. For microscopic traffic simulation there is no reason that the TAZs could not be as small as a single building if there is enough computing resource and the data available. The computing resource is cheap now, but unfortunately, the global single building population distribution database is not available yet. So TUMS uses LandScan as an alternative. There are 5657 LandScan Population Cells (LPC) comparing with 62 TAZs in Alexandria, VA. The LPC resolution is around 100 times higher than using TAZ.

The size of TAZ or LPC is related with the network level of details. A network with principle and minor arterials does not need high-resolution population dataset. In traditional traffic modeling the collect or local streets are ignored due to the low traffic volume. But if the OD zones use single buildings as the trip generation unit, then the network should include collect and local streets. For non-notice emergency evacuation simulation, the collect and local streets become very important because the evacuees who are close to the boundary of evacuation region can get out the evacuation area very quick by using local streets. If the local streets are excluded from the network, all these evacuees have to travel to the opposite direction in order to access the major arterials and then travel to the boundary points. This is unrealistic and generates artificial congestion on the arterials.

Unified Network and Population Database

There are many transportation modeling systems and traffic simulation tools available both on open source or in commercial packages, such as TRANSIMS,

MITSIM, VISSIM, SUMO and MATSIM, just to name a few. All of these models have similar basic input requirements such as network and population. But each one of them has its own input and output format. TUMS has developed a uniform database for the entire world and also has utilities to convert the unified database to different format for different models. Currently TUMS supports TRANSIMS and MITSIM. SUMO and MATSIM will be added in the near future.

Big Data

In modern era like todays, Data is generated by internet activity, sensors for environment, traffic cameras, satellite imagery and is referred to as Big Data. Processing, analyzing and visualizing this data have its own challenges. The OSM dataset has 3+ billion point and 300+ million ways (links) in the planet data file (December 2015 version). Among the 300+ million ways there are 80+ million street links. The LandScan has 93+ billion cells.

Since all these data are read only, TUMS chooses flat binary files to store the data for its simplicity and for easy random access. In order to retrieve the data efficiently, both the network and the population data are decomposed to 1 by 1 degree cells. The street network is stored in a shapefile format and the population data is store in a binary grid format. Since ESRI's binary grid format is a proprietary format, TUMS has to develop its own binary grid format. In TUMS database the OSM street network occupies 21 GB and LandScan occupied 4.6 GB disk space.

The vehicle trajectory data is another challenge. Assume that there are 100K vehicles in a median size county and the simulation time is 24 h for a daily commuter traffic simulation with 1-s simulation step, the total trajectory for all vehicles is 8+ billion points per scenario. This data is stored in ESRI-like point shapefile. Since ESRI's point shapefile has 2 GB file boundary limitation, TUMS developed its own binary format without the 2 GB boundary limitation. This vehicle trajectory data is streaming to TUMS web-based application for animation.

Urban Information System (UrbIS)

Leveraging Big Data to Understand Urban Impact on Environment and Climate

Cities are one of the major contributors to climate change. At the same time cities themselves are most strongly affected by the changing environmental conditions. Immense complexity of interactions between urban areas, climate, and natural environments presents scientists with a multitude of challenges. First, urban environments are characterized by a very large number of variables including demographics, energy, quality of the environment and many others. Second, cities show significant variety and strongly differ in terms of their processes and energy and material flux. Third, cities themselves have become major defining forces for their surrounding environments by affecting local topography, air circulation, water-heat balance and habitats. Resulting human-natural system has a large number of feedback loops and correlations among its variables.

Understanding of the urban environments can be improved by tapping into vast information resources that have become available to the researchers thanks to the Big Data technologies (Chowdhury et al. 2015). Traditional sources of Big Data like historical databases of Twitter messages, postings in other social networks, and cell phone locations can provide valuable insights into the functioning of people in urban environments. However, the majority of the data of interest for urban researchers exist in the form of an "ecosystem of small data", as a large number of disparate datasets created by different communities, government agencies, and research institutions. Finding such data and then merging them together for the use in a single analytical workflow has become a major hindrance for such studies. To address these challenges we at ORNL have embarked on the developing of ORNL Urban Information System (UrbIS)—a web-based software tool that would allow urban scientists to perform most of their analytical and data processing tasks in the cloud within a unified browser-based user interface.

UrbIS goal is to address a number of problems typically faced by the researchers in this area. After analyzing ORNL experience in a number projects including the ones described in this chapter we were able to identify multiple bottlenecks that impede scientists' productivity. These challenges can be mitigated by developing software for automating of several commonly performed tasks such as (1) finding the data necessary to achieve the goals of the study, (2) preparing the data from external sources for the use in the analytical software, (3) running modeling and analytical programs on the high-performance computing systems, and (4) retrieving and understanding the results of the analysis including representation of the results in visual form as graphs and maps.

Although scientists typically have a good understanding of the kinds of data they need for their research, finding specific datasets and not missing the relevant ones may be hard and time consuming. Most of the relevant data resides in the "deep web", i.e. not visible or not suitably indexed by general-purpose search engines like Google or DuckDuckGo. Therefore, such search engines often produce noisy results that require lots of manual filtering and verification or miss relevant data.

Search through dataset metadata provides a better alternative for finding scientific data. In the recent decade metadata has become a universally used tool for documenting large amount of data especially produced by the governmental, international, and other major research organizations. Multiple standardization efforts have generated several specifications that cover lots of aspects of important domain-specific knowledge necessary to precisely represents information about the data. Metadata search capabilities are currently available in many data archives and repositories such as, for example, NASA's Data Portal (https://data.nasa.gov/)

and DataONE Earth Observation Network (https://www.dataone.org/) supported by National Science Foundation.

Metadata search in most cases is more effectives than the use of general-purpose search engines because the metadata is structured and curated according to well defined standards. Users can filter through the data not only by the keywords or commonly used phrase but also by specific spatial, temporal or attribute information. For example, it is possible to limit the search by a specific sensor, variable, target area, time interval, or range of values. Certain results can be excluded from the search by using negation criteria that is not easily achievable in general-purpose search engines. However, typical metadata search requires interaction with multiple metadata search systems and familiarity with a variety of user interfaces and APIs.

After the necessary data have been identified the users have to extract relevant subsets of data (i.e. clipping a region of interest and/or limiting the data to a specific time interval) and move the data to their workstations. Sometimes the volume of the data can be very large like in the case of ensembles of global circulation models and can reach the volumes on the order of terabytes. Present-day hard drive costs are low enough not to be a limiting factor for storing data still movement of the large volumes of data over the network requires lots of time and special software like Globus Toolkit GridFTP[1]. The bigger problem is the maintenance of the harvested data on the workstation or local network storage that requires not only cataloguing of the data but also checking the dataset integrity, creating backups and retrieving updates for corrected errors or newer versions of the datasets.

The next preparation step is converting harvested data into the formats that can be understood by the analytical software. This step includes not only simple format conversion but also other non-analytical operations. Almost all of the urban data is spatiotemporal as the overwhelming majority of data records in urban datasets have some kind of geographic and time reference. Thus there is always a need to maintain and convert cartographic projection and other spatial referencing information. The datasets often come in the formats that are not understood by the analytical software or in-house developed code and scientists are forced to spend their time on developing format converters or perform lots of manual transformations. In case of UrbIS we are often faced with the data that comes from different scientific communities—urban scientists and climate modellers. Most climate and weather data is stored in NetCDF or HDF5 files while urban datasets mostly rely on the file formats of the commercial GIS software. Many of the open-source and commercial GIS are able to read these formats but the data have to be manually reorganized. The separate problem is semantics misalignment among the datasets especially when the datasets originate from different communities. This includes incompatibilities related to the units of measure, variable names, inconsistent naming of the grids and spatial regions. Such differences between the datasets are often not reflected in their metadata.

---

[1]http://toolkit.globus.org/toolkit/docs/latest-stable/gridftp/.

Many of the analytical and modeling projects at ORNL including the ones in this review heavily rely on high-performance computing systems and facilities. This includes conventional computing clusters, cloud-based systems and leadership massively parallel facilities like TITAN and Eos[2]. Developing, porting and using scientific code and managing applications and data on such systems require special technical skills and experience that are not commonly available.

Finally, the output of the high-performance models has to be presented in the form suitable for understanding and presentation. In our research domain this almost always means visualization in the geographic context with the help of advanced visualization tools found in the geographic information systems. At this point the modeling and analytical results should not only be converted into the formats understandable by GIS but also aligned with other pertinent geographic data.

Even though most of the outlined difficulties are technical in nature, they impose a significant toll on the scientists' time and increase overall costs of research. Moving these burdens from the scientist is one of the main goals of UrbIS. Earlier ORNL experience with similar systems has demonstrated efficiency of such approach. In 2010 ORNL has developed iGlobe—a desktop application for the geographic analysis of climate simulation data that combined server-side analysis and management of data with geographic visualization in a single workflow (Chandola et al. 2011). iGlobe is built around NASA WordWind Java[3] and allows users to retrieve the data from the data portals, process them on the server, and visualize analysis results on the desktop. Control of the server-side processing of the data is performed through desktop GUI using secure shell connection. Results of the analysis are presented using NASA WorldWind visualization component as interactive 2D or 3D geographic displays. With the advance of web, cloud and high-performance computing technologies we are leveraging our iGlobe experience at the new level of web-centric and cloud-centric applications.

Currently UrbIS is under active development and exists as an early evolving prototype available to ORNL internal users. When completed UrbIS will allow urban researchers to execute a complete analytical workflow starting with discovering and obtaining necessary data from diverse data repositories, analyzing them using high-performance computing capabilities, and then to visualizing and publishing the results. Researchers will be able to perform all these operations completely in the cloud and/or on the server through a standardized web interface. When fully implemented UrbIS will eliminate the need to download and process any input, intermediate, or output data files on the workstation.

Screenshots of the UrbIS prototype are presented from Figs. 9, 10, 11 and 12. UrbIS workflow starts with a federated metadata search interface (Fig. 9). This interface provides a user with the search capabilities through several external metadata search engines and internal ORNL data holdings. For the federated

---

[2]https://www.olcf.ornl.gov/titan/.

[3]http://worldwind.arc.nasa.gov/java/.

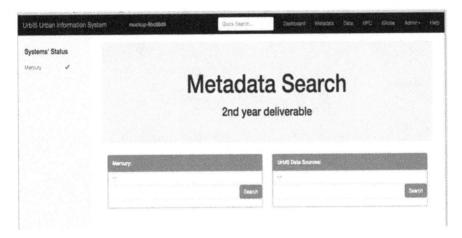

Fig. 9 UrbIS federated metadata search interface

Fig. 10 UrbIS workspace manager interface

metadata search engine we are using a customized version of Mercury[4]—an in-house ORNL metadata search engine that enables the search over other metadata repositories and archives like DataONE (https://www.dataone.org/) and ORNL DAAC (https://daac.ornl.gov/). In addition to the external data repositories UrbIS also provides its users with several frequently used datasets with common used

---

[4]http://mercury.ornl.gov/.

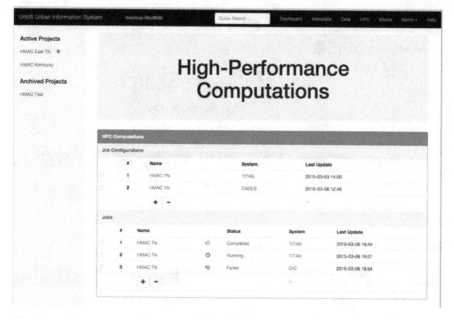

Fig. 11 HPC computations for analysis and modeling

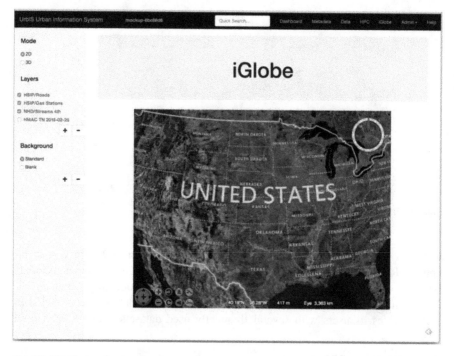

Fig. 12 UrbIS visualization interface

geographic information. All the datasets and datastores can be searched through the same interface completely transparent for the user.

After finding the needed data the user defines the region of interest and spatial resolution of his study area. At the same time in the background the system starts retrieval and sub-setting of the requested data from the external data stores. After the data has been placed into UrbIS scratch disk space the user will be notified and the records about downloaded subsets will appear in the workspace manager interface (Fig. 10). Here users can check the statistics of the downloaded data and verify completion of the download and conversion processes. All the retrieved data will be stored cloud-side and will not be downloaded to the user's workstation unless requested. Internally the data will be converted into application-specific representations optimized for further processing and access through UrbIS web services.

At the next stage of the workflow the user will be able to choose from a library of the analytical and modeling functionality (Fig. 11). As a part of the initial UrbIS development we are implementing high-performance clustering algorithms for building typologies of the cities based on a large number of input parameters. After specifying input parameters the user will submit a task to one of the high-performance computers. UrbIS will prepare the data in the form suitable for the selected processing method and create a batch configuration file containing commands for the target high-performance platform. The user will be able to initiate processing on the target system directly from UrbIS interface. After the job completion UrbIS will retrieve the results and convert them into the formats that are used internally.

The final step of the workflow is the visualization of the results in the geographic context. For that purpose we are using WebWorldWind (https://webworldwind. org/)—a modern javascript version of NASA WorldWind that utilizes WebGL. It can be launched from within popular browsers without the need to download any plugins or desktop applications. Visualization section of UrbIS (Fig. 12) has user interface typical for a digital globe like Google Earth or NASA WorldWind. Here the user can visualize the input and output data in the geographic context. The data is fed to the visualization component with WMS and WFS services from the internal UrbIS storage. Also the user can pull the data from any other data source supported by the NASA WebWorldWind including default WorldWind layers. The user will have an ability to switch between 3D and 2D views and choose the background and portrayal methods most suitable for his visualization purposes.

Current implementation of UrbIS is being developed using nodejs for the server side components. As a spatial data storage we are using PostgreSQL with PostGIS extensions. High-performance processing components are implemented as external modules and they use languages and tools most appropriate for the specific algorithms and platform. UrbIS should be accessible from any modern browser with WebGL support enabled (for visualization component). Internally UrbIS relies on service-oriented architecture with most functionality exposed through RESTful programming interface.

Currently UrbIS is in the active development and is available for testing to internal users. Its implementation will enable users to use high-performance and cloud-based infrastructure in their research and reduce the time needed for mundane tasks such data movement and format conversion. Also UrbIS will serve as a testing ground for new cloud-based technologies to facilitate the use of large geodata in scientific research within high-performance and cloud-based environment. After initial release and testing with internal user community we will proceed to implementing other sets of functionalities and extend the library of the high-performance analytical routines with other methods and models. In the future we plan to integrate UrbIS infrastructure with systems like Jupyter Notebooks (http://jupyter.org/) so that users can develop their own code through a web interface and access UrbIS data using web services.

Conclusions

Efforts to understand and analyze data-enabled science has created a clear need to unite various Earth Observation High-performance Computing (EO-HPC) systems, where the best of these various worlds are brought together in one shared Cyber-Infrastructure (CI) platform. In this chapter, we have discussed such a CI platform being developed at Oak Ridge National Laboratory using data-driven GeoComputation, novel analytical algorithms and emerging technologies. Systems interoperability, scalability and sustainability play an ever-increasing role in data-driven and informed decision-making process in our platform. We have discussed architectural and technical challenges in development of our platform, and broadening implications of it as illustrated by our research initiatives for data and science production. With technological roots in HPC, our platform is optimized for Earth Observation Big Data used to accelerate the research efforts, and foster knowledge discovery and dissemination more quickly and efficiently for US federal agencies.

Acknowledgements The authors would like to thank a number of US federal agencies for their continued support for the research presented here. Sincere gratitude is due to many of our Geographic Information Science and Technology group colleagues for their collaboration and assistance. This paper has been authored by employees the US Federal Government and of UT-Battelle, LLC, under contract DE-AC05-00OR22725 with the US Department of Energy. Accordingly, the US Government retains and the publisher, by accepting the article for publication, acknowledges that the US Government retains a non-exclusive, paid-up, irrevocable, worldwide license to publish or reproduce the published form of this manuscript, or allow others to do so, for US Government purposes.

References

Bhaduri B et al. (2002) LandScan. Geoinformatics 5(2):34–37
Bhaduri B et al. (2015a) Emerging trends in monitoring landscapes and energy infrastructures with big spatial data. SIGSPATIAL Spec 6(3):35–45

Bhaduri BL et al. (2015b) Monitoring landscape dynamics at global scale: emerging computational trends and successes. Oak Ridge National Laboratory, Oak Ridge, TN

Chandola V et al. (2011) iGlobe: an interactive visualization and analysis framework for geospatial data. Proceedings of the 2nd International Conference on Computing for Geospatial Research & Applications, 23 May 2011, p 21

Chowdhury P et al. (2015) An comparison of data storage technologies for remote sensing cyber-infrastructures. The International Conference on Big Data Analysis and Data Mining

Kalidindi SR (2015) Data science and cyberinfrastructure: critical enablers for accelerated development of hierarchical materials. Int Mater Rev 60(3):150–168

Karthik R (2014a) SAME4hpc: a promising approach in building a scalable and mobile environment for high-performance computing. Proceedings of the Third ACM SIGSPATIAL International Workshop on Mobile Geographic Information Systems, 4 November 2014, pp 68–71

Karthik R (2014b) Scaling an urban emergency evacuation framework: challenges and practices. Workshop on Big Data and Urban Informatics

OpenStreetMap (2016) https://www.openstreetmap.org. Accessed May 20 2016

Patlolla DR et al. (2012) Accelerating satellite image based large-scale settlement detection with GPU. Proceedings of the 1st ACM SIGSPATIAL International Workshop on Analytics for Big Geospatial Data, 6 November 2012, pp 43–51

Patlolla D et al. (2015) GPU accelerated textons and dense sift features for human settlement detection from high-resolution satellite imagery

Smith L et al. (1995) TRANSIMS: transportation analysis and simulation system. Los Alamos National Laboratory, New Mexico

Sorokine A et al. (2012) Tackling BigData: strategies for parallelizing and porting geoprocessing algorithms to high-performance computational environments. GIScience

12

Citizen Science for Observing and Understanding the Earth

Mordechai (Muki) Haklay, Suvodeep Mazumdar, and Jessica Wardlaw

Abstract Citizen Science, or the participation of non-professional scientists in a scientific project, has a long history—in many ways, the modern scientific revolution is thanks to the effort of citizen scientists. Like science itself, citizen science is influenced by technological and societal advances, such as the rapid increase in levels of education during the latter part of the twentieth century, or the very recent growth of the bidirectional social web (Web 2.0), cloud services and smartphones. These transitions have ushered in, over the past decade, a rapid growth in the involvement of many millions of people in data collection and analysis of information as part of scientific projects. This chapter provides an overview of the field of citizen science and its contribution to the observation of the Earth, often not through remote sensing but a much closer relationship with the local environment. The chapter suggests that, together with remote Earth Observations, citizen science can play a critical role in understanding and addressing local and global challenges.

Introduction

The term Earth Observation (EO) emerged in the late 1950s and 1960s to describe the use of space technology to observe and monitor natural and human-made phenomena across the globe. EO received an important boost in the concerted effort of scientists from many countries during the International Geophysical Year (IGY 1957–1958). In the period before the IGY, it was suggested that Earth-orbiting satellites will transform our understanding of the physical environment (e.g. Kaplan

The original version of this chapter was revised. An erratum to this chapter can be found at
https://doi.org/10.1007/978-3-319-65633-5_19

M. (Muki) Haklay (✉)
Extreme Citizen Science Group, UCL, London, UK
e-mail: m.haklay@ucl.ac.uk

S. Mazumdar
Department of Computing, Faculty of Arts, Computing, Engineering and Sciences, Sheffield
Hallam University, Sheffield, UK
e-mail: s.mazumdar@shu.ac.uk

J. Wardlaw
Faculty of Engineering, University of Nottingham, University Park, Nottingham, NG7 2RD
e-mail: jessica.wardlaw@nottingham.ac.uk

1956, p. 4). Indeed, the IGY saw the launch of Sputnik, the first human-made satellite, and the opening of the "Space Age". A decade later, in 1966, in light of rapid advances in space technology, an analysis of the potential of EO predicted a wide range of applications: from mapping parts of the world that were not yet mapped, to monitoring wildlife or forest fire, managing air pollution and many other benefits in the fields of geography, agriculture, water resources, oceanography, geology and archaeology (Willow Run Laboratories 1966). EO is the epitome of 'Big Science' (Ravetz 2006)—large-scale scientific endeavours, requiring complex and extremely expensive instruments, and meticulous planning and cooperation across the globe. Moreover, the data that arise from these efforts require specialist skills and tools (e.g. computers in the 1960s), which puts it beyond the abilities and financial reach of non-professional scientists. EO, in short, emerged when the role of the amateur scientist was diminishing, and the scope for members of the general public to participate meaningfully in cutting-edge scientific research became very limited.

And yet, the participation of non-professional scientists—people who have interest in scientific research but operate outside scientific institutions—is an integral part of the process of observing and understanding the Earth. Even in the early days of EO and the IGY in 1957, thousands of amateur scientists—nowadays called citizen scientists—participated in tracking these very early satellites (McCray 2006). Under the leadership of Fred Whipple, the then head of the Smithsonian Astrophysical Observatory, amateurs were engaged in identifying satellite locations in close collaboration with professional scientists. The Moonwatch project, which continued to run until 1975, involved participants in optical observation of satellites as they orbit the Earth. The programme faced obstacles and scepticism from other scientists and administrators of the IGY, as they did not trust the volunteers to provide sufficiently high quality information and observations. Eventually, though, it was a group of Moonwatch volunteers who first observed the Sputnik (McCray 2006). In many ways, the story of Moonwatch is mirrored in current citizen science projects that are the focus of this chapter.

The 1966 analysis (Willow Run Laboratories 1966) is striking for its emphasis on aspects of EO that are now taken for granted: a set of instruments, with predictable characteristics, taking many readings over large areas of the Earth in a consistent way, and enabling scientific analysis at scales that could not exist before. The ability to capture information over large areas automatically is presented in contrast to the difficulties of large-scale observations on the ground: for example, through a network of meteorological observations by volunteers and professional observers that can potentially be replaced by satellite measurements. Eventually, meteorology demonstrates, though, that the satellite has not replaced the observers on the ground completely. Although the area of meteorology relies on a vast array of automated sensing systems, satellite observations and a large and well-funded professional observation network, there is still a role for volunteers who can report information from their homes. More generally, instead of an assumption that a set of automated instruments will eventually replace the contribution of observers on the ground and that the non-professional scientist will eventually be consigned to history, a more nuanced picture has emerged, in which citizen scientists, professional scientists and EO experts are working collaboratively to address local and global challenges.

Societal and technological changes underlie this transition in the area of EO, and the next section will explore the trends that have led to the current incarnation of citizen science. Following this, we provide an overview of the main areas that are covered by citizen science, using a typology that shows how both old and new approaches to citizen science shaped the current landscape. The typology does not cover the full range of citizen science activities, but rather highlights sample activities along three axes: domains of scientific activities, reliance on digital technologies and level of engagement of participants in shaping projects. We also use the typology to introduce a range of projects that demonstrate each type of citizen science activity area within the area of Earth Observations. Following these examples, we suggest some of the critical technical and societal aspects of citizen science in the context of EO.

Societal and Technological Trends

While public participation in scientific efforts has the long history that was noted above, the final decade of the twentieth century marked the beginning of a new form, which received the name *citizen science*. The first recorded use of the term is in 1989 and describes how 225 volunteers across the USA collected rain samples to assist the Audubon Society, a nature conservation organisation, in an acid-rain awareness-raising campaign (Kerson 1989). The volunteers collected samples, checked for acidity and reported back to the organisation, therefore creating coverage across the continent and demonstrating the extent of the acid-rain phenomenon.

The term continued to gain recognition slowly throughout the 1990s (e.g. Bonney 1996). The past decade has seen a rapid increase in the number of citizen science projects and their scale. As a result, citizen science is now the accepted term for a range of practices. The term was first noted in Wikipedia in 2005 and recognised by the Oxford English Dictionary in 2014 as *"scientific work undertaken by members of the general public, often in collaboration with or under the direction of professional scientists and scientific institutions"* (OED 2014).

In addition to the term citizen science, this form of public involvement in scientific research has also been termed Public Participation in Scientific Research (PPSR), participatory science, civic science and amateur science, as well as crowdsourced science. In specific areas of scientific research, citizen scientists are known by domain-specific terms such as birdwatchers, amateur astronomers, volunteer weather observers or amateur archaeologists. This variety points to the longevity of the practice and the current convergence under an umbrella term due to the growing importance of these practices.

As was noted in the introduction, there are many parallels between the early days of citizen science participation in EO and its current incarnation. Yet, when examining the current scale and depth of engagement, current citizen science has clearly moved beyond previous forms of public involvement in scientific research.

Several societal and technological trends help to explain the emergence of citizen science today. These include the rapid growth in (especially higher) education during the second part of the twentieth century, increased leisure time (especially in middle and high income countries), and growth in educated and able retirees. On the technical side, the growth of the Web and mobile communication, and the ubiquitous connectivity that they offer, is highly significant. In particular, the emergence of Web 2.0 and the evolution of peer-production systems in the past 10 years, as well as the development and proliferation of cheap sensors that can collect data from the environment, played an important role. We now turn to look briefly at each of these.

The second half of the twentieth century has seen a major transformation in education across the world, with countries such as the UK moving from 1.6% of the population with a tertiary level of education in 1950, to 21.7% in 2010 (Barro and Lee 2012). This translates into a rise from <1 in 50 to over 1 in 5 in the span of 60 years. More generally, across advanced economies, the rate rose from 2.8% in 1950 to 17.9% in 2010. Importantly, this transition happened while the size of the population itself increased almost twofold, and significant improvements occurred at all levels of education, as both culture and education across the developed world became more oriented towards scientific thinking (Flynn 2007). This education shift has provided many millions of people across the world with the cognitive ability to understand abstract concepts, logic and hypothetical ideas. Education levels continue to increase across the world with an estimated 240 million people studying for tertiary education in 2013, of which about 2.5 million are studying at doctoral level (UIS 2015). Of course, not every person in tertiary education studies Science, Technology, Engineering and Mathematics (STEM) topics and yet the number of people who can potentially participate in citizen science without intensive training in the principles of the scientific method, or in the need for accurate measurements and the necessity to follow data collection protocols rigorously, is very large and will continue to increase.

In conjunction with the rise in levels of scientific education, the time dedicated to work in advanced economies decreased during the late twentieth century, down to about 40 h across OECD countries. The introduction of a 2-day weekend across advanced economies during the middle part of the century freed up time for leisure, hobbies, volunteering activities and family time.

The final societal aspect of note in the context of this discussion is increased life expectancy, which, combined with slow changes in retirement age, has led to a growth in educated and healthy people in their 60s and 70s who are active in their communities. For some, citizen science provides a way of re-engaging with topics of science that they studied earlier in their life, but have not engaged with during their working career.

On the technical side, the main factors are more familiar and have been covered extensively in the media and academic literature (e.g. Cuff et al. 2008; Haklay et al. 2008; Haklay 2013). Especially within the context of EO, we can identify several important trends. First is the growth of the Web and the ability to access scientific information through platforms from Wikipedia or lectures recorded on YouTube, to scientific papers that are shared through repositories and Open Access journals. Moreover, the Web is not only a conduit to consume scientific knowledge but also a suitable medium for creating new ways of engaging very large groups of people (which are referred to as the Crowd) to perform tasks and shared activities. Frequently described as user-generated content, the Web ushered in a new form of interaction between people with limited technical capacity and web-based systems, in which they could share news and information about their locality with a potentially global audience. From the point of view of EO, the ability of participants to generate geographic information and share it is especially important, and this was recognised by Mike Goodchild in 2007 as Volunteered Geographic Information or VGI (Goodchild 2007). Another aspect of this new mode of interaction is the ability to engage thousands and even millions of participants in performing small tasks that, in aggregate, yield significant results, as well as involving a very large group of participants in solving problems (this is known as crowdsourcing).

The ability to generate VGI is also linked to the removal of the selective availability of the Global Positioning System (GPS) signal in May 2000, and the subsequent proliferation of GPS receivers and location-based technologies (Haklay et al. 2008). The provision of an easy-to-use and automated location tracking and recording mechanism that can be easily communicated with latitude and longitude coordinates attached as metadata to different items of information (geotagging) is highly significant to citizens' participation in EO. With the advance of VGI, geotagged images from people's smartphones are available in quantities and at temporal and spatial scales that were never seen before and, more importantly, shared in a machine-readable way (e.g. Antoniou et al. 2010).

However, while GPS receivers are vital ingredients of EO, many other sensors also reduced in size and cost, due to the proliferation of smartphones with computing and sensing abilities in the past decade. It is now common to have multiple sensors in a smartphone, including a barometer, camera, microphone, accelerometer, electronic compass and more, which are integrated into the device to enable its functioning (e.g. the ability to acquire location rapidly in the case of the barometer) but can be reused by a range of applications to perform scientific measurements.

Next, we should also note the growth in internet bandwidth both at home and through mobile telecommunication networks. At home, the ability to send and receive videos and large image files, as well as rapid and responsive interaction with websites, is critical to many citizen science projects. Moreover, the possibility to stay connected while on the move increases the volunteers' ability to record and share observations quickly and easily: sometimes as small tasks that last a few seconds (micro-tasks) or even by carrying the device itself passively.

Finally, and most recently, there has been a growth in Do-It-Yourself (DIY) electronics with the introduction of easy-to-programme control boards or minia-turised computers (such as Arduino or Raspberry Pi), 3D printers allowing rapid prototyping and small-scale manufacturing, and hubs such as Makers clubs and Hackspaces where people meet and work together to develop new devices and projects. The falling cost of sensors and components that was mentioned above, and the practice of sharing of information over collaborative websites, opened up the ability of mostly technically savvy people to carry out their own DIY science efforts.

The trends have fundamentally altered citizen science. Most of the public in the early twentieth century could not be relied upon to identify and report the scientific names of species (though some expert amateur naturalist has done so) and were not equipped with scientific understanding; nor were they carrying around powerful scientific instruments in their pockets. In contrast, today, hundreds of millions of people have such abilities, and therefore the potential for participation is much higher. Yet, it is important to note how the multiple underlying trends are also defining the demographics of those who participate in citizen science. Participants in citizen science activities are typically well educated, working in a job that provides enough income and working conditions for ample leisure, and have access to the Internet as well as own a smartphone. Not surprisingly, because of the imbalances in care responsibilities, science education and income, men are overrepresented in citizen science. For example, a study found that 87% of the participants in a volunteer computing project (see the next section) were men (Krebs 2010), while a similar bias was identified in ecological observations of birds (Cooper and Smith 2010). Internationally, citizen science is concentrated in advanced economies, especially the USA and northern Europe. The need to access the Internet still presents an obstacle, with level of access ranging from 87% in the UK, to 81% in the USA, and only 65% in European countries such as Poland or Portugal (ITU 2013). At the more local level, even for those who have access to a smartphone, many of the software applications (apps) that support citizen science assume continuous and seamless Web connectivity, even though 3G and 4G coverage is partial in highly urbanised environments such as London or New York City, let alone in remote nature reserves. Language can also present a barrier. As the background material and the apps are being developed by scientists, the amount of discipline-specific jargon and the level of understanding that is needed to get involved in a project can exclude many people. Finally, since English is the main language of scientific papers and of science more generally, many of the tools and technologies that support citizen science activities rely on knowledge of English, and are not available in local languages, especially in areas of high cultural heterogeneity such as Europe.

The result is somewhat ironic. Much of the rhetoric of citizen science is about its potential for inclusion of new groups in society, raising awareness and interest in the scientific enterprise, and providing new routes for education and skills. The current demographics demonstrate that, without purposeful effort, this will not happen. Sometimes, there are simple routes to overcoming challenge (e.g. to provide paper forms in areas of low connectivity) but, more generally, special attention should be paid to those that are, mostly unintentionally, excluded from citizen science activities.

Citizen Science Today: Main Areas of Activity

The aforementioned trends provide an explanation of the current integration of citizen science within EO activities. In this section, we will look at a general typology of major types of activities in citizen science, which are identified by their domain, technical needs and the level of engagement of participants in the projects.

Figure 1 presents the topics that the following sections will cover. Under the banner of citizen science we can see three types of activities, each highlighting a different facet of the field. First, *long-running citizen science* is defined by activities that involve the public in areas where the practice of working in collaboration with non-professional scientists is well established. There are many areas of science in which volunteers continue to play a role in research and, from the perspective of EO, *ecological and biological observations*, *weather observations* and participation in *archaeology* provide good examples of the potential of citizen science. Other areas, such as astronomy, have also demonstrated sustained engagement with citizen science. The next type of citizen science projects highlights the way technology influences citizen science, and includes projects that rely on the Internet and the Web. These *citizen cyberscience* projects use the ability of computers as both computing and communication devices to engage citizen scientists. In fact, projects that fall under this category would not have existed without the proliferation of computers and the Internet. Here we find *volunteer computing*, which utilises the unused computing resources of participants' computers; *volunteer thinking*, which asks the participants to contribute through their cognitive abilities; and *passive sensing*, which relies on the sensors that are integrated into mobile computing devices to carry out automatic sensing tasks. The final group of citizen science projects that will be discussed here emphasises the depth of engagement of participants, and we will term these as *community science*—projects that are carried out as part of local, everyday settings, to address local concerns and needs. Here we look at three types of activities: *participatory sensing*, a joint activity between researchers and members of the public with varied levels of participation in setting what will be

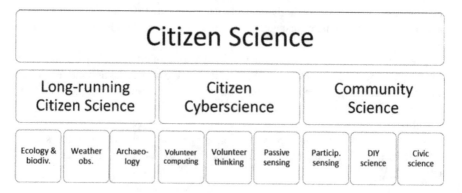

Fig. 1 Mapping current citizen science activities by domains, technology and engagement

detected, where and how it will be analysed; *DIY science*, in which participants create the scientific instrument themselves, and repurpose a range of materials and tools to build laboratories and carry out their enquiries; and, finally, *civic science*, which covers scientific activities that aim to build relationships between the public, experts and policy makers and enable them all to participate in scientific knowledge production (Bäckstrand 2003). This type of scientific practice can also be recognised as bottom-up science (McQuillan 2014).

To understand how each of these families of citizen science operates and their relevance to EO, the following sections will look at each, in turn.

Citizen Science Across Domains: Long-Running Citizen Science

At first sight, there are many areas of scientific activities that continued to engage with non-professional scientists throughout the era of Big Science: birdwatching in biological and ecological observations (Kobori et al. 2016, Bonney et al. 2009), recording of meteorological conditions (WMO 2001), and volunteers in archaeological digs (Clarke 1978) are all examples of sustained engagement of citizen scientists. However, the aforementioned trends have changed the interaction with volunteers and the way in which they carry out their work and, especially, share information.

For example, new technologies are making a step change in the relevance of volunteered ecological and biodiversity observations for wider EO systems. Historically, amateur naturalists (as they were known) recorded information in their notebooks, frequently using idiosyncratic records management systems, and the sharing of the information with others was partial. As August et al. (2015) discuss, the use of digital technologies not only supports the immediate sharing of information, but also contributes in a structured way: for example through predefined forms on websites and increased use of apps on smartphones, which provide further information such as GPS coordinates, geolocated images or audio recordings (Jepson and Ladle 2015, Powney and Isaac 2015). In some of these systems and apps, information can appear in global databases (e.g. the Global Biodiversity Information Facility—GBIF) instantaneously. Therefore, through the link between more educated volunteers and ICT-enabled streamlined sharing, current citizen science contributes to the creation of EO systems in the area of biodiversity.

As noted, the participation of volunteers in weather and meteorological obser-vations is also well documented. The network of meteorological observations is one of the longest-standing examples of citizen science, with many thousands of volunteers reporting local meteorological conditions to national organisations, which improves the quality of modelling and understanding weather and climate (WMO 2001). As such, this area demonstrates a union between citizen science and established professional science that is both persistent and evolving over time through the development of instruments and the abilities of participants. For example, the commercial provider of meteorological forecasts, The Weather Company, is managing a large-scale crowdsourcing aggregation of weather observations through the Weather Underground network. A network of over 180,000 participants link observations from their personal weather stations to improve The Weather Company's predictions, and benefit by receiving personalised forecasts. Another example of the scale and scope of citizen science in this area is provided by the UK Met Office Weather Observation Website (wow.metoffice.gov.uk), which received 38 million observations in its first year of operation in 2011 (POST 2014), and provides a source of additional information for the Met Office that is especially useful during extreme weather events. Here, too, technological advances streamlined and standardised information sharing, while the increased awareness and skills in the general public contributed to greater participation in reporting.

Archaeology is another field with a tradition of voluntary participation with historical links to EO. Looking back at the 1966 report (Willow Run Laboratories 1966), satellites were seen as an extension of aerial photography, which was already in use at the time in archaeology. However, while satellite instruments were expected to assist in identifying large features, such as buried cities, "... *the requirements for the use of such sensors in the detection of small features remain very near and possibly beyond the capabilities of orbital sensor equipment as presently envisioned*" (p. 153). Today, there is a flourishing sub-discipline of Space Archaeology, which uses the abilities of EO to advance the field. Citizen science, in the form of crowdsourcing, now addresses the exact problem that, 50 years ago, was considered beyond the possible. In 2010, Albert Yu-Min Lin and colleagues devised a system based on high resolution satellite imagery to engage over 10,000 volunteers in the task of assessing potential locations for the unknown burial site of Genghis Khan (Lin et al. 2014). The system asked volunteers to evaluate an area visually and mark locations that they considered as potentially interesting. The ability to engage a huge number of volunteers enabled the examination of a very large area (6000 km^2), yielding 55 candidate sites for further archaeological studies on the ground. The application that was developed for this task eventually evolved into the Tomnod system, now used by Digital Globe for humanitarian and other crowdsourcing efforts. Here, the ability of people to collaborate online is significant, and vividly demonstrates the importance of broadband and the bidirectional web in opening up new avenues for collaboration between professional and non-professional researchers.

The Impact of Technology: Citizen Cyberscience

As the overview noted, the emergence of the Internet and the Web as a global infrastructure has enabled a new incarnation of citizen science, which has been termed citizen cyberscience by Francois Grey (2009) and could not possibly have existed before. Characteristically it relies on the proliferation of billions of connected personal computing devices—desktop computers, smartphones and games consoles—and utilises the computational and sensing power of these devices to double as scientific instruments. If the previous section considered how citizen science is integrated into different scientific disciplines, here we look at how advances in personal computing transformed the potential of citizen science in contributing to EO. In particular, we will focus on three subcategories: *volunteered computing, volunteered thinking* and *passive sensing*.

Volunteered computing was first launched in 1999, with the SETI@home project (Anderson et al. 2002), which exploits the unused processing capacity in personal computers and uses the Internet to send and receive work packages that are analysed automatically and sent back to the main server. The system on which SETI@home is based, the Berkeley Open Infrastructure for Network Computing (BOINC), is now used for over 100 projects. While volunteer computing is popular in the area of biological and medical research, it is not well utilised in the area of EO. An example of the potential of volunteer computing is provided by the ClimatePrediction.net project, which was established by climate researchers at the University of Oxford in 2002 and, with exposure from mass media, reached 60,000 volunteers. In the early months of 2014, when the project team wanted to suggest the degree to which recent floods could be attributed to climate change, they were able to run over 33,000 different models and demonstrate that it is highly likely that the floods were more severe due to climate change (Climateprediction.net 2014).

While volunteered computing asks very little from the participants, apart from installing software on their computers, *volunteered thinking* engages volunteers at a more active and cognitive level (Grey 2009). In these projects, participants use a website in which information or an image is presented to them. They are provided with a little training in the task of classifying the information, after which they are exposed to information that has not been analysed and are asked to carry out classification work. Galaxy Zoo (Lintott et al. 2008) is one of the most well-known and developed examples of volunteer thinking. Over 100,000 volunteers classified images of galaxies for this project, and it spawned a range of applications that are included in the wider Zooniverse set of projects (see http://www.zooniverse.org/). We have already encountered one example of volunteer thinking work in the previous section, with the effort to locate Genghis Khan's tomb.

Another highly relevant example of the involvement of volunteer thinking in EO is provided through the OpenStreetMap project (Haklay and Weber 2008). This distributed project has now engaged millions of people in mapping their area through a combination of tracing satellite imagery and on the ground survey, as demonstrated by the "Missing Maps" project (Feinmann 2014) in which areas that were not mapped before are being added to OpenStreetMap to support humanitarian efforts.

The final example of citizen cyberscience is provided by *passive sensing*, in which participants either connect sensors to their computers or smartphones, or use the built-in sensors that are available in devices, to support EO efforts. Unlike participatory sensing, which we will encounter in the next section, passive sensing is mostly based on automatic data capture and sharing, without the conscious intervention of the volunteer. We have seen one example of such passive sensing in the Weather Underground network above. The personal weather stations that are linked to the network operate automatically, mostly without intervention from their owner, and, once they are set to deliver the information to The Weather Company's server, they will continue to do so. However, further potential for EO integration is provided via mobile devices. For example, the Quake Catcher Network (QCN) is utilising the movement sensors that are integrated into some laptop computers, to enhance observations from existing seismic observation stations (Cochran et al. 2009). QCN is improving the quality of seismic information that is emerging from events. Interestingly, QCN is utilising the BOINC framework but extends it by linking to sensors.

Depth of Participation: Community Science

Community science is a term used here to describe citizen science projects with a significant element of bottom-up control over the project; at its extreme, activities are initiated and driven by a group of participants who identify a problem that is a concern for them and address it using scientific methods and tools. The problem definition, data collection and analysis might be carried out by community members or in collaboration with scientists in established laboratories whose role is to support and carry out work on behalf of the community members. This is in contrast to the types of citizen science discussed above, where the scientific research question, data collection methodology and the analysis are all done by professional scientists and the role of participants is somewhat restricted.

In the area of community science, three examples demonstrate the role of participants and professional scientists, and their potential of integration with EO.

First, *participatory sensing* is defined as sensing activity in which a group of participants contribute together to a body of information. Importantly, while the term is now used liberally to describe a wide range of crowdsourced sensing activities with varying levels of active engagement with the citizen scientists who will carry out the sensing, in the original definition (Burke et al. 2006; Goldman et al. 2009), *"Participatory Sensing emphasizes the involvement of citizens and community groups in the process of sensing and documenting where they live, work, and play ..."* (p. 4). Unlike passive sensing, the participants are expected to take a more significant role in shaping the sensing project. In its simple form, participatory sensing requires lower cognitive effort from participants and relies on users to provide sensory information in a structured manner via their mobile devices and cloud services (Estrin 2010). The participants select when and where to carry out data collection, but the application and the data infrastructure are set. The examples that were mentioned above of apps for ecological and biodiversity recording operate under this scheme—many of the apps that are provided to volunteers (see Jepson and Ladle 2015, Powney and Isaac 2015) expect participants to take an image of the species that they have identified using their smartphones and share them by adding them to national or global databases. Another interesting example is Ikarus (http://thermal.kk7.ch/), where paraglider flight log data is collected and processed to generate thermal maps. With a large number of paragliders and flight paths, Ikarus is one of the largest participatory sensing initiatives (Von Kaenel et al. 2011).

In contrast, the practices of *DIY science* mean that the participants develop instruments, methodologies for data collection and analysis (Nascimento et al. 2014). This requires very deep engagement from the participants, as well as technical and scientific knowledge to carry out the scientific study in question. In the area of EO, we can see an emerging interest in the development of devices and software that can facilitate balloon and kite mapping, for example by the Public Laboratory for Open Technology and Science (Public Lab for short). By using simple adapted technology, digital cameras are strapped to balloons or kites and used to observe and analyse local conditions. Simple adaptation can convert a camera to near-infrared, and thus provide information at other wavelengths than visible light (Breen et al. 2015). Moreover, if the group who collected the data wishes, this very detailed local mosaic can be shared through Google Earth.

The final type of community science is *civic science*, which is explicitly linked to community goals and questions the state of things. While some DIY science is done from such a perspective (e.g. Breen et al. 2015), civic science can also include work with indigenous communities in the use of smartphones to record community resources and other local features, even when the participants are non-literate (Stevens et al. 2014). While the approach is highly sensitive to local cultural practices and involves a lengthy discussion about information sharing to ensure consent, it can be integrated into larger EO systems, providing the unique perspective of local and traditional ecological knowledge.

Citizen Science and Earth Observation: Technical, Societal, Ethical and Policy Aspects

Based on this overview of citizen science and understanding of the areas that it covers, the main methods that are used in it and their linkage to EO systems, we now turn to common issues that are discussed in respect of citizen science and EO. Here we focus on technical, societal and ethical aspects, and the policy issues that can facilitate or hinder this integration.

Technical Aspects

An ever-increasing number of citizen science or crowdsourcing initiatives continue to engage user communities in scientific endeavours, incorporating a variety of mechanisms for collecting, presenting and analysing crowdsourced information— some of them with hundreds of thousands of participants (e.g. Lintott et al. 2008;

Fishwick 2014). While the engagement of large numbers of volunteers is always encouraging, increased volumes of data do not necessarily imply the presence of more useful information (Mackechnie et al. 2011). Data collected by citizens may lack metadata regarding their quality, which can often lead to being discredited by many scientists (Alabri and Hunter 2010). Although many studies (e.g. Newman et al. 2003; Foster-Smith and Evans 2003; Fore et al. 2001) have found volunteers can collect comparable data to professional researchers (with limitations), there is a need for a methodological approach towards addressing quality concerns in citizen science data; even a traditional consensus-based approach to volunteered data is non-trivial to apply (Salk et al. 2017). It is only expected that citizen science and crowdsourced data, owing to its increasing relevance and contribution to scientific outcomes, should also adhere to similar strict and rigorous validation processes in the specific fields, for example in meteorology or biodiversity observations, as we have seen above. This calls for a need to have a good understanding of the validation mechanisms that can be potentially applied to citizen science initiatives to mitigate against the potential economic (Foody 2015) and policy implications that we will see later.

Haklay (2016) identified six mechanisms that citizen science initiatives typically employ for quality assurance: some rely on the principle of abundance where validity of information is based on agreement by other observers (crowdsourcing) or other volunteers more experienced than the original observer (social). Validation mechanisms can also involve either the presence of geographical or domain expertise. Several initiatives rely on measuring instruments to provide their quality, precision and accuracy (instrumental observation), which can, in turn, indicate the validity of observations. Finally, the more formal process oriented relies on trained participants for collecting and storing observations in a highly structured manner, using standard equipment. One or a combination of validation mechanisms is typically employed in citizen science initiatives. For example, the COBWEB (Higgins et al. 2016) and WeSenseIt projects employ multiple levels of validation such as position accuracy (instrumental observation), linked data and various forms of automated data validation (domain). The recently concluded Crowd4Sat project's (see http://www.crowd4sat.eu/) demonstration projects individually employ a few validation mechanisms such as crowdsourcing, geographical and instrumental observation.

Citizen science can also be used to improve EO data. For example, EO data is prone to errors and inconsistencies arising from the very nature of its sensing. For example, an intrinsic problem with measuring water or snow coverage in mountainous regions with Synthetic Aperture Radar (SAR) imaging is the slant-range distortion effect. This leads to inaccurate assessments of snow and water cover. Providing mechanisms for special interest communities such as hikers, wildlife enthusiasts and photography groups who frequent such areas can provide critical fine-grained spatial data to help rectify such issues.

Another issue is that, while satellites can provide critical information during emergency events, they typically have revisit times that are too low to address urgent issues. Flooding, for instance, is an emergency scenario where situations can drastically deteriorate in a very short time. In such scenarios, passive and opportunistic participatory sensing information can improve the temporal resolution of information critical to emergency responders (Endsley 1995) and thus improve data availability and quality.

Another critical aspect of the integration of citizen science data and EO is information integration. In EO, the potential for understanding and predicting natural and human-made processes and phenomena is enormous. However, it involves observing a variety of key variables (e.g. local information, sensor observations, human perception of events and phenomena). Often, the variables of interest are the ones that are most difficult to observe such as flow and velocity, land occupation, etc. However, decision makers can only achieve a holistic view once all the variables are contextualised in the same model and view. Data fusion can provide enriched information from multiple sensor data in a variety of granularities (El Faouzi et al. 2011, Koch 2014) by exploiting spatio-temporal features of data. This is a highly complicated task since different variables represent different characteristics, in different contexts and levels of granularity. One of the findings from the Crowd4Sat project observed that, while satellite observations are on a large scale, human observations are typically on a much smaller scale and, as a result, several discrepancies can arise when trying to use either observation to validate the other. While most citizen science initiatives, in their current state, are independent, effective data fusion and integration techniques promise a more integrated approach, where different citizen science initiatives can inform and provide support for each other as well as share information. The Crowd4Sat project identified this as one of the possible directions where citizen science and crowdsourcing could evolve in the next few years, and potentially change the landscape of citizen science for EO.

Societal and Ethical Aspects

Citizen science and EO integration is not only a technical issue, but a socio-technical system that requires human and societal perspectives, which need attention if the EO field is to continue extending the activities outlined above in a sustainable manner. Many of these are addressed by the European Citizen Science Association's 10 Principles of Citizen Science (ECSA 2015). These translate into three themes of practical importance to which we now turn: motivations, ethics and privacy.

First, the voluntary participation of citizens is implicit to the practice of citizen science, and this demands attention regarding their motivation to engage and sustain their engagement over time as this could demand financial, societal or other forms of recognition (Geoghegan et al. 2016). As citizen cyberscience activities demonstrate, different activities require different levels of engagement from volunteers. Demands on volunteers' time and resources are associated with a risk of disengagement over the course of the activity on the one hand, and potential exploitation and abuse by those leading the project on the other.

Citizen science implies elements of engagement in genuine scientific activity and education. In line with standard ethical practices, we can argue that citizen scientists should be treated with respect; informed as to the purpose of their involvement or if this is not possible, and they have been deliberately misled, debriefed; and involved with only non-harmful activities. Another stance is to treat participants as collaborators and not waste their time (Prestopnik and Crowston 2012; ECSA 2015). Participants' informed consent and transparency regarding the storage and use of data are also desirable to mitigate against the increasing privacy concerns discussed next.

In citizen cyberscience and participatory sensing especially, terms such as Human Processing Units, Remote Person Calls and The Human API can all be found in the literature (Reeves 2013; Lease and Alonso 2014), which can be interpreted in a way that dehumanises participants. This may be to the detriment of citizen science projects, as it can harm motivations that we noted above. The fundamental implication of this is clear: citizen science requires consideration of the needs and requirements of public participants.

Within the ethical consideration, we should pay special attention to privacy issues. These are given comprehensive coverage by Bowser et al. (2014) and include a range of data ownership and sharing challenges, not least issues of intellectual property. Advancements in citizens' participation in EO will, however, need a balanced approach towards not only ethics, but privacy and associated legalities. A key concern in fusing crowdsourced data is in understanding the implications of cross-sensor/cross-project/cross-initiative data fusion. For example, in the recently started Big Data project, Seta (see http://setamobility.eu/) has highlighted this as a key concern. The project, aimed at understanding mobility in metropolitan areas, attempts to integrate different forms of motorised and non-motorised crowdsourced data. While anonymisation policies are designed to protect privacy of citizens, integrating data across multiple sensors and facets may pose a significant risk to their identification.

Citizen Science Integration into Policy

The final aspect of citizen science and EO that we will consider here is the integration of citizen science into policy. This is critical for the wider discussion on open science and this book as, without long-term integration, citizen science will remain in niche activities and, as in the early period of EO, will be treated with suspicion or ignored, and hence will not receive the necessary financial and organisational resources that will enable it to thrive.

One aspect of citizen science that is raising challenges for policy makers is the multiple outcomes that these projects can achieve and the domains to which they contribute. As we have seen, and as many noted (Bonney 1996; Burke et al. 2006; Bonney et al. 2009; Haklay 2015), citizen science can contribute to increasing awareness of participants to science and environmental issues. The training, which can lead to an improvement of data quality, can also assist in gaining skills that are not directly linked to the specific project. Citizen science activities such as DIY workshops or ecological surveys can act as science outreach activities, while at the same time teaching participants the value of sharing information. Such multiple goals sometimes mean that there is no single owner or funder for such projects within regular organisational structures, and therefore is a mixed blessing for the field.

However, the scale and reach of citizen science and the visibility of projects over the past decade has raised the attention of policy makers at local, regional, national and international level. Such awareness, and the development of appropriate policies as well as long-term funding mechanisms, is critical to the sustainability of citizen science efforts and ensuring that the information is being used in the long run.

For EO, probably the most significant demonstration of the integration of citizen science and EO is within the Eye on Earth Alliance, which brings together the Abu-Dhabi Global Environmental Data Initiative (AGEDI), the Group on Earth Observations (GEO), the International Union for the Conservation of Nature (IUCN), the United Nations Environment Programme (UNEP) and the World Resources Institute (WRI). The alliance committed at the end of 2015 to promoting the use of citizen science as an integral part of the Sustainable Development Goals (SDG) monitoring activities, in a way that integrates it with EO. There are emerging examples of the use of citizen science in systems such as UNEP Live or the WRI Global Forest Watch. However, these examples are still in an experimental stage, and a transition to fully fledged systems is still in the future. As noted before, and likely due to the urgency associated with it, the one area in which crowdsourcing, citizen science and EO are already being integrated and used is in the humanitarian response to disasters where systems such as OpenStreetMap (Zook et al. 2010) or Tomnod are now routinely used.

Conclusions

In this chapter, we have introduced the area of citizen science and its relationship with EO. As we have seen, while citizen science has interacted with EO from its very early days, the new incarnation of citizen science provides scale, scope and coverage that transform it into a new component with true global reach. After 50 years of EO, and while there are many challenges ahead, citizen science is evolving into a pivotal provider of information about the planet. Traditional areas of citizen science activities are enhanced by current technological and societal activities and the information that is provided by citizen scientists can now be verified and tested, and therefore be integrated into EO systems faster. New areas, forms of engagement and capabilities also emerged recently, and these also contribute, either through passive or participatory sensing, to the range of activities that can be included in citizen science.

Yet, as we have seen in the discussion, the long-standing challenges of data quality, data integration, interoperability, management of metadata, engagement, interaction, privacy and ethics are all significant to the process of improving citizen science outcomes and ensuring that it will become a sustainable practice, and that the information that is emerging from it is used in many areas of policy and decision making.

Acknowledgement This chapter is partially based on material that was published in Haklay (2013) and Haklay (2015), which was updated, extended and edited for this chapter. The research was kindly supported by: the UK's Engineering and Physical Sciences Research Council (awards EP/I025278/1, EP/K022377/1); the European Union's Seventh Framework Programme (FP7/2007-2013) under grant agreements EveryAware (no 265432), Citizen Cyberlab, WeSenseIt (no 308429), and iMars (no 607379); European Union Horizon 2020 projects DITOs, Seta, ECSAnVis; and the European Space Agency's Crowd4Sat. We would also like to thank Berris Charnley for useful comments and suggestions for an earlier version of the manuscript.

References

Alabri A, Hunter J (2010) Enhancing the quality and trust of citizen science data. In IEEE Sixth International Conference on e-Science. IEEE, Washington, DC, pp 88–81

Anderson DP, Cobb J, Korpela E, Lebofsky M, Werthimer D (2002) SETI@home: an experiment in public-resource computing. Commun ACM 45(11):56–61

Antoniou B, Haklay M, Morley J (2010) Web 2.0 geotagged photos: assessing the spatial dimension of the phenomenon. Geomatica 64(1):99–110

August T, Harvey M, Lightfoot P, Kilbey D, Papadopoulos T, Jepson P (2015) Emerging technologies for biological recording. Biol J Linn Soc 115(3):731–749

Bäckstrand K (2003) Civic science for sustainability: reframing the role of experts, policy-makers and citizens in environmental governance. Glob Environ Polit 3(4):24–41

Barro R, Lee J-W (2012) A new data set of educational attainment in the World, 1950–2010, NBER Working Papers 15902. National Bureau of Economic Research, Cambridge, MA

Bonney R (1996) Citizen Science: a lab tradition. Liv Bird 15(4):7–15

Bonney R, Ballard H, Jordan R, McCallie E, Phillips T, Shirk J, Wilderman CC (2009) Public participation in scientific research: defining the field and assessing its potential for informal science education. A CAISE inquiry group report. Center for Advancement of Informal Science Education (CAISE), Washington, DC

Bowser A, Wiggins A, Shalney L, Preece J, Henderson S (2014) Sharing data while protecting privacy in citizen science. Interactions 21(1):70–73

Breen J, Dosemagen S, Warren J, Lippincott M (2015) Mapping grassroots: geodata and the structure of community-led open environmental science. Int E J Critic Geograph 14(3):849–873

Burke JA, Estrin D, Hansen M, Parker A, Ramanathan N, Reddy S, Srivastava MB (2006) Participatory sensing. Center for Embedded Network Sensing, Los Angeles, CA

Clarke DV (1978) Excavation and volunteers: a cautionary tale. World Archaeol 10(1):63–70

ClimatePrediction.net (2014) Weather@home 2014: the causes of the UK winter floods. http://www.climateprediction.net/weatherathome/weatherhome-2014/. Accessed Aug 2014

Cochran ES, Lawrence JF, Christensen C, Jakka RS (2009) The quake-catcher network: citizen science expanding seismic horizons. Seismol Res Lett 80(1):26–30

Cooper CB, Smith JA (2010) Gender patterns in bird-related recreation in the USA and UK. Ecol Soc 15(4):4. http://www.ecologyandsociety.org/vol15/iss4/art4/. Accessed Aug 2014

Cuff D, Hansen M, Kang J (2008) Urban sensing: out of the wood. Commun ACM 51(3):24–33

El Faouzi N-E, Leung H, Kurian A (2011) Data fusion in intelligent transportation systems: progress and challenges – a survey. Informa Fus J 12(1):4–10

Endsley MR (1995) Toward a theory of situation awareness in dynamic systems. Hum Fact 37(1):32–64

Estrin D (2010) Participatory sensing: applications and architecture [internet predictions]. IEEE Int Comput 14(1):12–42

European Citizen Science Association (ECSA) (2015) ECSA ten principles of citizen science. http://ecsa.citizen-science.net/sites/ecsa.citizen-science.net/files/ECSA_Ten_principles_of_citizen_science.pdf. Accessed Apr 2016

Feinmann J (2014) How MSF is mapping the world's medical emergency zones. BMJ 349:7540

Fishwick C (2014) Tomnod–the online search party looking for Malaysian Airlines flight MH370. Guardian, 14

Flynn JR (2007) What is intelligence? Beyond the Flynn effect. Cambridge University Press, Cambridge

Foody G (2015) Valuing map validation: the need for rigorous land cover map accuracy assessment in economic valuations of ecosystem services. Ecol Econ 11:23–28

Fore LS, Paulsen K, O'Laughlin K (2001) Assessing the performance of volunteers in monitoring streams. Freshwater Biol 46:109–123

Foster-Smith J, Evans SM (2003) The value of marine ecological data collected by volunteers. Biol Conserv 113:199–213

Geoghegan H, Dyke A, Pateman R, West S Everett G (2016) Understanding motivations for citizen science. Final report on behalf of UKEOF, University of Reading, Stockholm Environment Institute (University of York) and University of the West of England

Goldman J, Shilton K, Burke J, Estrin D, Hansen M, Ramanathan N, Reddy S, Samanta V, Srivastava M, West R (2009) Participatory sensing: a citizen-powered approach to illuminating the patterns that shape our world. Foresight & Governance Project, White Paper, pp 1–15

Goodchild M (2007) Citizens as sensors: the world of volunteered geography. Geo J 69:211–221

Grey F (2009) The age of citizen cyberscience, CERN Courier, 29th April 2009. http://cerncourier.com/cws/article/cern/38718. Accessed Jul 2011

Haklay M, Weber P (2008) Openstreetmap: user-generated street maps. IEEE Pervas Comput 7(4):12–18

Haklay M (2013) Citizen science and volunteered geographic information – overview and typology of participation. In: Sui DZ, Elwood S, Goodchild MF (eds) Crowdsourcing geographic knowledge. Springer, Berlin, pp 105–122

Haklay M (2015) Citizen science and policy: a European perspective. The Woodrow Wilson Center, Commons Lab, Washington, DC

Haklay M (2016) Volunteered geographic information, quality assurance. In: Richardson D, Castree N, Goodchild M, Liu W, Kobayashi A, Marston R (eds) The international encyclopedia of geography: people, the earth, environment, and technology. Wiley/AAG, Hoboken, NJ

Haklay M, Singleton A, Parker C (2008) Web mapping 2.0: the Neogeography of the Geoweb. Geography Compass 3:2011–2039. https://doi.org/10.1111/j.1749-8198.2008.00167.x

Higgins C, Williams J, Leibovici D, Simonis I, Davis M, Muldoon C, Genuchten P, O'Hare G (2016) Citizen OBservatory WEB (COBWEB): a generic infrastructure platform to facilitate the collection of citizen science data for environmental monitoring - work in progress. Int J Spat Data Infrastruct Res 11:20

International Telecommunications Union (ITU) 2013. Percentage of Individuals using the Internet 2000–2012, ITU (Geneva), June 2013. http://www.itu.int/en/ITU-D/Statistics/Documents/statistics/2013/Individuals_Internet_2000-2012.xls. Accessed Mar 2016

Jepson P, Ladle RJ (2015) Nature apps: waiting for the revolution. Ambio 44(8):827–832

Kaplan J (1956) The International Geophysical Year Program of the United States. In: Crary AP, Gould LM, Hulburt EO, Odishaw H, Smith WE (eds) Antarctica in the international geophysical year: based on a symposium on the Antarctic. American Geophysical Union, Washington, DC. https://doi.org/10.1029/GM001p0001

Kerson R (1989) Lab for the Environment. MIT Technol Rev 92(1):11–12

Kobori H, Dickinson JL, Washitani I, Sakurai R, Amano T, Komatsu N, Kitamura W, Takagawa S, Koyama K, Ogawara T, Miller-Rushing AJ (2016) Citizen science: a new approach to advance ecology, education, and conservation. Ecol Res 31(1):1–19

Koch W (2014) Tracking and sensor data fusion. Springer, New York, NY. https://doi.org/10.1007/978-3-642-39271-9

Krebs V (2010) Motivations of cybervolunteers in an applied distributed computing environment: MalariaControl.net as an example. First Monday, 15(2)–1 February 2010

Lease M, Alonso O (2014) Crowdsourcing and human computation, Introduction. Encyclopedia of social network analysis and mining (ESNAM), pp 304–315. http://link.springer.com/referenceworkentry/10.1007/978-1-4614-6170-8_107. Accessed Apr 2016

Lin AYM, Huynh A, Lanckriet G, Barrington L (2014) Crowdsourcing the unknown: the satellite search for Genghis Khan. PLoS One 9(12):e114046

Lintott CJ, Schawinski K, Slosar A, Land K, Bamford S, Thomas D, Raddick MJ, Nichol RC, Szalay A, Andreescu D, Murray P, van den Berg J (2008) Galaxy zoo: morphologies derived from visual inspection of galaxies from the Sloan Digital Sky Survey. Month Not Roy Astronomic Soc 389(3):1179–1189

Mackechnie C, Maskell L, Norton L, Roy D (2011) The role of 'Big Society' in monitoring the state of the natural environment. J Environ Monit 13:2687

McCray WP (2006) Amateur scientists, the International Geophysical year, and the ambitions of Fred Whipple. Isis 97(4):634–658

McQuillan D (2014) The countercultural potential of citizen science. M/C J 17(6). http://journal.media-culture.org.au/index.php/mcjournal/article/view/919. Accessed May 2016

Nascimento S, Guimarães Pereira A, Ghezzi A (2014) From citizen science to do it yourself science: an annotated account of an on-going movement. Joint Research Centre/European Commission, London

Newman C, Buesching CD, MacDonald DW (2003) Validating mammal monitoring methods and assessing the performance of volunteers in wildlife conservation – "Sed quis custodiet ipsos custodies?". Biol Conserv 113:189–197

Oxford English Dictionary (OED) (2014) Citizen science. Accessed Aug 2014

Parliamentary Office for Science and Technology (POST) (2014) POSTNote 476 Environmental Citizen Science, UK Parliament

Powney GD, Isaac NJ (2015) Beyond maps: a review of the applications of biological records. Biol J Linn Soc 115(3):532–542

Prestopnik N, Crowston A (2012) Purposeful gaming & socio-computational systems: a citizen science design case. In Proceedings of GROUP'12, October 27–31, 2012, Sanibel Island, Florida, USA, pp 75–84.

Ravetz J (2006) The no-nonsense guide to science. New Internationalist

Reeves S (2013) Human-computer interaction issues in human computation. In: Michelucci P (ed) Human computation handbook. Springer, New York, pp 411–419

Salk CF, Sturn T, See L, Fritz S (2017) Limitations of majority agreement in crowdsourced image interpretation. Trans GIS 21:207. https://doi.org/10.1111/tgis.12194

Stevens M, Vitos M, Altenbuchner J, Conquest G, Lewis J, Haklay M (2014) Taking participatory citizen science to extremes. Pervas Comput IEEE 13(2):20–29

UNESCO Institute of Statistics (UIS) (2015) Enrolment by level of education, December 2015 release. UNESCO, Paris

Von Kaenel M, Sommer P, Wattenhofer R (2011) Ikarus: large-scale participatory sensing at high altitudes. In Proceedings of the 12th Workshop on Mobile Computing Systems and Applications. ACM, New York, NY, pp 63–68

Willow Run Laboratories (1966) Peaceful uses of earth-observation spacecraft: Volume II: Survey of applications and benefits. Report 7219-1-F (II), Ann Arbor, Michigan, February 1966

World Meteorological Organisation (2001) Volunteers for weather, climate and water. Geneva, Switzerland, WMO No. 919

Zook M, Graham M, Shelton T, Gorman S (2010) Volunteered geographic information and crowdsourcing disaster relief: a case study of the Haitian earthquake. World Med Health Pol 2(2):7–33

Permissions

All chapters in this book were first published by Springer; hereby published with permission under the Creative Commons Attribution License or equivalent. Every chapter published in this book has been scrutinized by our experts. Their significance has been extensively debated. The topics covered herein carry significant findings which will fuel the growth of the discipline. They may even be implemented as practical applications or may be referred to as a beginning point for another development.

The contributors of this book come from diverse backgrounds, making this book a truly international effort. This book will bring forth new frontiers with its revolutionizing research information and detailed analysis of the nascent developments around the world.

We would like to thank all the contributing authors for lending their expertise to make the book truly unique. They have played a crucial role in the development of this book. Without their invaluable contributions this book wouldn't have been possible. They have made vital efforts to compile up to date information on the varied aspects of this subject to make this book a valuable addition to the collection of many professionals and students.

This book was conceptualized with the vision of imparting up-to-date information and advanced data in this field. To ensure the same, a matchless editorial board was set up. Every individual on the board went through rigorous rounds of assessment to prove their worth. After which they invested a large part of their time researching and compiling the most relevant data for our readers.

The editorial board has been involved in producing this book since its inception. They have spent rigorous hours researching and exploring the diverse topics which have resulted in the successful publishing of this book. They have passed on their knowledge of decades through this book. To expedite this challenging task, the publisher supported the team at every step. A small team of assistant editors was also appointed to further simplify the editing procedure and attain best results for the readers.

Apart from the editorial board, the designing team has also invested a significant amount of their time in understanding the subject and creating the most relevant covers. They scrutinized every image to scout for the most suitable representation of the subject and create an appropriate cover for the book.

The publishing team has been an ardent support to the editorial, designing and production team. Their endless efforts to recruit the best for this project, has resulted in the accomplishment of this book. They are a veteran in the field of academics and their pool of knowledge is as vast as their experience in printing. Their expertise and guidance has proved useful at every step. Their uncompromising quality standards have made this book an exceptional effort. Their encouragement from time to time has been an inspiration for everyone.

The publisher and the editorial board hope that this book will prove to be a valuable piece of knowledge for researchers, students, practitioners and scholars across the globe.

List of Contributors

Luigi Ceccaroni
1000001 Labs, Barcelona, Spain

Filip Velickovski and Laia Subirats
Eurecat, Barcelona, Spain

Meinte Blaas and Anouk Blauw
Deltares, Delft, The Netherlands

Marcel R. Wernand
Royal Netherlands Institute for Sea
Research, Texel, The Netherlands

Heike Bach
Remote Sensing in Geosciences,
VISTA GmbH, Munich, Germany

Wolfram Mauser
Department of Geography, University
of Munich LMU, Munich, Germany

David J. Lary, Gebreab K. Zewdie,
Xun Liu, Daji Wu, Estelle Levetin,
Rebecca J. Allee, Nabin Malakar,
Annette Walker, Hamse Mussa,
Antonio Mannino and Dirk Aurin
Hanson Center for Space Sciences,
The University of Texas at Dallas, 800
West Campbell Road, Richardson,
TX 75080, USA

Ravi Kapur
Imperative Space, London, UK

Val Byfield
National Oceanography Centre,
Southampton, UK

Fabio Del Frate
University of Tor Vergata, Rome,
Italy

Mark Higgins
EUMETSAT, Darmstadt, Germany

Sheila Jagannathan
World Bank Group, Washington, DC,
USA

Bessie Schwarz, Beth Tellman,
Jonathan Sullivan, Catherine Kuhn,
Richa Mahtta and Laura Hammett
Cloud to Street, USA

Gabriel Pestre
The Data Pop Alliance, New York,
USA

Bhartendu Pandey
Yale University, New Haven, CT,
USA

Diego Scardaci
EGI Foundation & INFN Catania
Division, Amsterdam, The
Netherlands

Rajasekar Karthik, Alexandre
Sorokine, Dilip R. Patlolla,
Cheng Liu, Shweta M. Gupte and
Budhendra L. Bhaduri
Geographic Information Science
and Technology Group, Oak Ridge
National Laboratory, Oak Ridge, TN,
USA

Maria Antonia Brovelli, Marco
Minghini and Giorgio Zamboni
Department of Civil and Environmental
Engineering, Politecnico di Milano,
Piazza Leonardo da Vinci 32, 20133,
Milano, Italy

Conor O'Sullivan and Nicholas Wise
Satellite Applications Catapult, Harwell, UK

Pierre-Philippe Mathieu
ESA/ESRIN, Frascati, Italy

Peter Baumann, Vlad Merticariu and Dimitar Misev
rasdaman GmbH, Bremen, Germany
Jacobs University, Bremen, Germany

Angelo Pio Rossi, Brennan Bell, Ramiro Marco Figuera and Huu Bang Pham
Jacobs University, Bremen, Germany

Oliver Clements
Plymouth Marine Laboratory, Plymouth, UK

Ben Evans
National Computational Infrastructure (NCI), Australian National University, Canberra, ACT, Australia

Heike Hoenig
rasdaman GmbH, Bremen, Germany

Patrick Hogan
NASA Ames, Moffett Field, CA, USA

George Kakaletris and Panagiota Koltsida
CITE s.a, Attiki, Greece

Simone Mantovani
MEEO s.r.l., Ferrara, Italy

Stephan Siemen and Julia Wagemann
ECMWF, Reading, UK

Andreea Bucur, Stefano Elefante and Vahid Naeimi
Department of Geodesy and Geoinformation, Technische Universität Wien, Gußhausstraße 27-29/E120, 1040, Vienna, Austria

Wolfgang Wagner
Department of Geodesy and Geoinformation, Technische Universität Wien, Gußhausstraße 27-29/E120, 1040, Vienna, Austria
EODC Earth Observation Data Centre for Water Resources Monitoring GmbH, Gusshausstrasse 27-29/CA 02 06, 1040, Vienna, Austria

Christian Briese
EODC Earth Observation Data Centre for Water Resources Monitoring GmbH, Gusshausstrasse 27-29/CA 02 06, 1040, Vienna, Austria

Tiziana Ferrari and Sergio Andreozzi
EGI Foundation, Amsterdam, The Netherlands

Mordechai (Muki) Haklay
Extreme Citizen Science Group, UCL, London, UK

Suvodeep Mazumdar
Department of Computing, Faculty of Arts, Computing, Engineering and Sciences, Sheffield Hallam University, Sheffield, UK

Jessica Wardlaw
Faculty of Engineering, University of Nottingham, University Park, Nottingham, NG7 2RD, UK

Index

Printed in the USA
CPSIA information can be obtained
at www.ICGtesting.com
JSHW011400091023
49903JS00004B/32